AF753643

The ASQ Certified Six Sigma Green Belt Study Guide

The ASQ Certified Six Sigma Green Belt Study Guide

Second Edition

Roderick A. Munro, Daniel J. Zrymiak, and Elizabeth J. Rice, Editors

ASQExcellence
Milwaukee, Wisconsin

© 2024 by ASQExcellence

Publisher's Cataloging-in-Publication Data

Names: Munro, Roderick A., 1955- editor. | Zrymiak, Daniel J., 1969- editor. | Rice-Munro, Elizabeth J., 1951- editor.
Title: The ASQ certified Six Sigma Green Belt study guide , second edition / Roderick A. Munro, Daniel J. Zrymiak, and Elizabeth J. Rice, Editors
Description: Includes bibliographical references. | Milwaukee, WI: ASQExcellence, 2024.
Identifiers: LCCN: 2024942715 | ISBN: 978-1636941073 (paperback) | 978-1636941080 (pdf) | 978-1636941097 (epub)
Subjects: LCSH Six sigma (Quality control standard)—Examinations—Study guides. | Quality control—Management—Examinations—Study guides. | Production management—Examinations—Study guides. | BISAC BUSINESS & ECONOMICS / Quality Control | STUDY AIDS / Professional
Classification: LCC TS156.17.S59 2024 | DDC 658.5/620218—dc23

No part of this book may be reproduced in any form or by any means, electronic, mechanical, photocopying, recording, or otherwise, without the prior written permission of the publisher.

ASQ and ASQExcellence advance individual, organizational, and community excellence worldwide through learning, quality improvement, and knowledge exchange.

Bookstores, wholesalers, schools, libraries, businesses, and organizations: Quality Press and ASQExcellence books are available at quantity discounts for bulk purchases for business, trade, or educational uses. For more information, please contact Quality Press at 800-248-1946 or books@asq.org.

To place orders or browse the selection of all Quality Press and ASQExcellence titles, visit our website at: http://www.asq.org/quality-press.

27 26 LS 7 6 5 4 3

ASQ
Quality Press
600 N. Plankinton Ave.
Milwaukee, WI 53203-2914
Email: books@asq.org
Excellence Through Quality™

ASQE™
Inspire Excellence

Table of Contents

Introduction

Welcome to the second edition of *The ASQ Certified Six Sigma Green Belt Study Guide*! This study guide presents over 500 questions with solutions, using *The ASQ Certified Six Sigma Green Belt Handbook, Third Edition* as the primary basis for the following practice questions. In Part A, questions and answers are arranged in order of the major sections of the Green Belt Body of Knowledge. Part B presents a 100-question simulated practice exam in which the questions are presented in random order.

Let's address the most obvious item first: Why does one need a second edition? The ASQ Certified Six Sigma Green Belt (CSSGB) Body of Knowledge (BoK) has been updated and, therefore, has become an expected level of knowledge for all who become Certified Six Sigma Green Belts. Who updates the BoK? The Six Sigma Green Belt Certification Committee specializes in keeping the BoK and all ASQ textbooks and materials current. It meets regularly and comprises highly qualified ASQ volunteers from around the world.

The primary purpose of this study guide is to use the questions as a self-assessment tool to help determine your readiness to sit for the ASQ certification exam.

The primary user of this material is the individual who plans to prepare to sit for the ASQ CSSGB examination. *The ASQ Certified Six Sigma Green Belt Study Guide* is arranged in the following manner:

1. There is one right answer per question. In some cases, caution is needed for formulas related to rounding errors.
2. The solutions appear immediately following the questions for each section.
3. The correct answers are indicated by letter and the corresponding section of the BoK. visit https://asq.org/cert/six-sigma-green-belt.
4. If the student did not answer correctly and believes additional preparation would be useful, it is recommended that they refer to the *Handbook* as a starting point for a refresher and deeper dive into the material, and even additional reference books, such as topic-specific references offered by ASQ.

Please know that this study guide cannot be taken into ASQ exam rooms during the actual exam process. The ASQ Certification Committee uses a well-honed exam-question-writing process that is designed to produce a thorough competency examination of the exam-taker.

Your feedback and suggestions for improvement to *The ASQ Certified Six Sigma Green Belt Study Guide, Second Edition* are most welcome. Please contact us at books@asq.org so that we can continually improve our resources and incorporate suggestions for improvement into future printings or editions of the book.

—Roderick A. Munro, Daniel J. Zrymiak, and Elizabeth J. Rice

Acknowledgments

I would like to thank *The ASQ Certified Six Sigma Green Belt Handbook* team with Govind Ramu and Daniel J. Zrymiak for making the SSGB process so interesting over the past 15-plus years with a callout to Tom Kubiak and James Bossert for their ongoing support of our team. Also, special thanks to Elizabeth J. Rice for her dedication to the quality profession and for being a loving partner over the past 30 years. I very much appreciate all of you.

—Roderick A. Munro

This book is dedicated to Susan, Brooke, Ryan, my family, friends, and the community of like-minded ASQ fellows, lifelong learners, and practitioners. I also dedicate this to my superior and mentor, Alex Doaga, and the quality innovators at Westgate Technology.

I would like to thank Tom Kubiak, Rod Munro, Elizabeth Rice-Munro, and Govind Ramu for the enduring impact and remarkable legacy of ASQ's Six Sigma handbooks and study guides.

I am personally grateful to many, particularly authors and editors, including Denis Leonard, Jayet Moon, Ron Stephens, Sandeep Kumar, Jim Bossert, Gregory Watson, and Jd Marhevko, for their collaboration and synergies over the years. I also extend my gratitude to ASQ's Quality Management Division (particularly Denis Devos and Peggy Milz) and the other Technical Communities, along with the Six Sigma practitioners worldwide, for their indulgence and pursuit of professional growth and excellence.

As we continue our journey from rough origins toward wisdom, virtue, and perfection, may we always proceed with curiosity, enthusiasm, and perseverance toward fulfilling manifestations of quality realized in our lives.

—Daniel J. Zrymiak

With love to my mother, Helen Frances Rice (the first female Naval Intelligence officer in WWII), and my father, Francis E. Rice (Marine Corps officer, WWII, Pacific), whose mutual passion for faith, country, and lifelong learning resulted in a multitude of awe-inspiring accomplishments. They were, and though gone from this earth, continue to be never-ending amazing sources of inspiration.

—Elizabeth J. Rice

CONTRIBUTORS' ACKNOWLEDGMENTS

The American Society for Quality and Quality Press would like to thank the Quality Press Peer Review Committee for its invaluable volunteer participation and contributions to this work. Without our volunteers' subject matter expertise, time, and passion for creating content, none of our efforts would be possible.

2023-2026 COMMITTEE MEMBERS

Scott A. Laman, Chair
Jane Keathley, Vice Chair
Melvin Alexander
Ahmad Elshennawy
Marc Hamilton
Gary Jing
Trevor Jordan
Peter Pylipow
G.S. Sureshchandar
Theia Theurer
Mary McShane-Vaughn

Part A

Sample Questions by BoK

Section I	Six Sigma and the Organization
Section II	Definite Phase
Section III	Measure Phase
Section IV	Analyze Phase
Section V	Improve Phase
Section VI	Control Phase

Section I
Six Sigma and the Organization

(Exam Targets 11 Questions)

A. Six sigma and organizational goals

1. Value of six sigma

Recognize why organizations use six sigma, how they apply its philosophy and goals, and the evolution of six sigma from quality leaders such as Juran, Deming, Shewhart, Ishikawa, and others. (Understand)

2. Organizational goals and six sigma projects

Identify the linkages and supports that need to be established between a selected six sigma project and the organization's goals, including SMART goals, and describe how process inputs, outputs, and feedback at all levels can influence the organization as a whole. (Understand)

3. Organizational drivers and metrics

Recognize key business drivers (profit, market share, customer satisfaction, efficiency, product differentiation, and key performance indicators (KPIs)) for all types of organizations. Understand how key metrics and scorecards are developed and how they impact the entire organization. (Understand)

B. Lean principles in the organization

1. Lean concepts

Define and describe lean concepts such as the theory of constraints, value chain, flow, takt time, just-in-time (JIT), Gemba, spaghetti diagrams, and perfection. (Apply)

2. Value-stream mapping

Use value-stream mapping to identify value-added processes and the steps or processes that produce waste, including excess inventory, unused space, test inspection, rework, transportation, and storage. (Understand)

C. Design for six sigma (DFSS) methodologies

1. Road maps for DFSS

Distinguish between DMADV (define, measure, analyze, design, verify) and IDOV (identify, design, optimize, verify), and recognize how they align with DMAIC. Describe how these methodologies are used for improving the end product or process during the design (DFSS) phase. Understand how verification and validation are used to compare results against stated goals. (Understand)

2. Basic failure mode and effects analysis (FMEA)

Use FMEA to evaluate a process or product and determine what might cause it to fail and the effects that failure could have. Identify and use scale criteria, calculate the risk priority number (RPN), and analyze the results. (Analyze)

3. Design FMEA and process FMEA

Define and distinguish between these two uses of FMEA. (Apply)

QUESTIONS

1. Why is Six Sigma important in a business application?

   a. Customers are demanding the use of Six Sigma in many industries.

   b. Organizations must stay profitable to remain in business, and Six Sigma is a powerful tool to that end.

   c. Customers demand that all suppliers use Six Sigma to continually improve products.

   d. Managers demand that Green Belts be trained to support organizations in today's businesses.

2. What is the main goal of the Six Sigma philosophy?

   a. Reduced variation

   b. Reduced waste

   c. Increased production

   d. Increased profits

3. What is the most common cause of a failed Six Sigma program in an organization?

   a. Failure to properly define the processes to be utilized in the organization

   b. Lack of proper training for the various team members

c. Too many organizational definitions of what Six Sigma should be

d. Management's lack of commitment to real process improvement

4. The person who first developed the concept of what we call statistical process control (SPC) was:

a. W. Edwards Deming.

b. Joseph M. Juran.

c. Kaoru Ishikawa.

d. Walter A. Shewhart.

5. The cause-and-effect, or fishbone, diagram was created by:

a. W. Edwards Deming.

b. Joseph M. Juran.

c. Kaoru Ishikawa.

d. Genichi Taguchi.

6. The quality guru who wrote extensively about the three key managerial processes (quality planning, quality control, and quality improvement) was:

a. W. Edwards Deming.

b. Joseph M. Juran.

c. Kaoru Ishikawa.

d. Walter A. Shewhart.

7. Which of the following tools is a process approach used for improving projects?

a. Red *X*

b. Pareto principle

c. Plan-do-check-act (PDCA)

d. Input-process-output (IPO)

8. Which element of SMART goals ensures that the goal is realistic and attainable?

a. Specific

b. Measurable

c. Achievable

d. Time-bound

9. SMART goals have become increasingly popular for what use?
    a. Measuring process control
    b. Ensuring only appropriate goals are met
    c. Assessing an organization's position before deciding on any new strategy
    d. Assessing project management
10. SMART goals are also known as:
    a. objectives.
    b. purposes.
    c. events.
    d. directions.
11. Which definition of KPI is most often used by Green Belts?
    a. Key Performance Indicator
    b. Key Preference Indicator
    c. Key Process Indicator
    d. Key Presentation Indicator
12. Which of the following statements describes a KPI?
    a. It is a quantifiable measurement, agreed to in advance, that reflects critical success factors.
    b. It is a specific measure, agreed to in the first-year strategy, that drives the success of the indicator.
    c. It is a strategic goal, agreed to in the five-year plan, that improves the bottom line.
    d. It is a surreptitious process that is periodically agreed upon to ensure continual improvement.
13. Sorting through all the data being collected and identifying the key measures to track are key challenges for any organization. Who is responsible for guiding the measures tracked in projects?
    a. The CEO and CFO must determine and direct the strategic measures to be tracked in projects.
    b. The management team gives direction on what measures are tracked in projects.

c. The Green Belt gives instruction on the most environmentally important project measurements to track.

d. The Master Black Belt and Executive Team identify and prioritize each measure to be tracked and the desired organizational impact.

14. Which of the following actions is a crucial consideration when selecting an organization's key performance indicators (KPIs)?

a. Choosing KPIs that only reflect financial performance

b. Selecting KPIs that are easy to measure but unrelated to goals

c. Aligning KPIs with the organization's strategic objectives

d. Prioritizing KPIs that can be measured in isolation from other metrics

15. The balanced scorecard includes measures from which perspectives?

a. Financial, production, human resources, and field service

b. Inventory, employee learning and growth, production counts, and quality

c. Financial, employee learning and growth, customer, and internal processes

d. Strategy, waste, customer, and the value stream

16. Business performance measures that are used to provide an approach to measuring multiple aspects of a business, including the financial aspect, are:

a. A2 reports.

b. statistical process control.

c. failure mode and effect analyses.

d. balanced scorecards.

17. Which of the following three concepts are fundamental to the understanding of lean thinking?

a. Purpose, waste, and waste elimination

b. Prevention, value, and continual improvement

c. Value, waste, and the process of creating value without waste

d. Purpose, pull, and partnering without interference

18. Which of the following concepts is associated with the Toyota Production System House?

a. Value for the customer

b. Cost of poor quality

c. Ishikawa diagram

d. Taguchi loss function

19. Which of the following actions is a principle of "The Toyota Way"?

a. Building a culture of stopping to fix problems to get quality right the first time

b. Encouraging employees to use the A3 methodology to sell their Six Sigma ideas to their Champions

c. Allowing employees to go-see and report their understandings back to management so that it stays out of the way

d. Teaching the management team to manage by wandering around

20. Studies have shown that an overwhelming percentage of lead time is non-value added, yet, over the years, efforts to decrease lead time have often focused on:

a. accelerating value-added functions.

b. reducing or eliminating non-value-added functions.

c. developing new manufacturing facilities with faster production lines.

d. decreasing the prototype design phase.

21. 5S is a workplace organization method that:

a. can eliminate the defect count found in many processes.

b. can fix prior mistakes in the organization process.

c. ensures a reduction of more than 80 percent of efficiency issues.

d. can help improve the efficiency and management of operations.

22. The 5S step of sort means:

a. label unused items for the archives.

b. remove unneeded items.

c. clean up any mess.

d. find a place for each item.

23. A team working in a lab cleans the hot plates and testing equipment in the steam baths. Which phase of 5S is it likely in?

a. Sort

b. Sustain

c. Shine

d. Standardize

24. The 5S step of standardize involves:

    a. creating checklists, standards, and work instructions to keep the work area in a clean and orderly condition.

    b. writing work procedures, posting OSHA standards, and writing flowcharts.

    c. creating strategic plans and delivering all-employee monthly standardized broadcasts about workplace cleanliness.

    d. creating work instructions and posting them in all areas with cleaning supplies.

25. The 5S step of sustain involves:

    a. training and testing of all the 5S steps.

    b. management support and employee empowerment.

    c. additional training in accident reduction.

    d. engaging management in the cost reductions effected with 5S.

26. The theory of constraints is:

    a. a problem-solving methodology applied to reduce energy usage.

    b. a problem-solving methodology that focuses on the weakest link in a chain of processes.

    c. a decision-making tool that focuses on prioritization of weak links.

    d. an analysis tool that uncovers covert processes.

27. The ultimate goal of lean is:

    a. reduced waste.

    b. customer satisfaction.

    c. perfection.

    d. one-piece flow.

28. The series of activities that an organization performs, such as ordering, designing, producing, and delivering products and services, is known as:

    a. an open system.

    b. a multiple-path production stream.

    c. a value stream.

    d. a transformation concept.

29. The symbol below is used on a value stream map to denote which of the following items?

    a. Supermarket

    b. Production kanban

    c. Customer

    d. Safety or buffer stock

30. On a value-stream map, inspections would be classified as:

    a. non-value added.

    b. value added.

    c. voice of the customer.

    d. kanban.

31. Unused space, defects, and rework are examples of:

    a. the cost of quality.

    b. waste.

    c. constraints.

    d. the value chain.

32. The method that sets the pace of industrial manufacturing lines so that production cycle times can be matched to customer demand rates is called:

    a. takt.

    b. a pull system.

    c. one-piece flow.

    d. single-minute exchange of dies (SMED).

33. A customer requires 320 units per day. The plant operates continuously for one eight-hour shift. What is the required takt time in minutes per unit?

    a. 0.67

    b. 1.5

c. 40

d. 320

34. Which of the following tools can be used during the improvement phase of the DMAIC model to evaluate workflow?

    a. Cause-and-effect diagram

    b. Root cause analysis

    c. Gemba walk

    d. Takt time analysis

35. Just-in-time (JIT) manufacturing can be described as:

    a. a storage setting with little or no manufacturing material inventory on hand at the site.

    b. the rate at which you need to complete a product to meet customer demand.

    c. a holding place of manufacturing material inventory with little or no incoming inspection.

    d. an optimal material requirement planning system for a manufacturing process.

36. Which of the following methods comprise the two pillars of the Toyota Production System (TPS)?

    a. Variable change and TRIZ

    b. Just-in-time and jidoka

    c. Kaizen and trend

    d. Validation and jidoka

37. A spaghetti diagram is best described as a:

    a. graphic chart that illustrates the pace of product flow.

    b. graphic tool intended to show the flow or path that people, products, or paperwork take during normal operations.

    c. chart showing the critical path for people and products.

    d. messy roadmap that shows flow or paths of multiple employees, materials, and paperwork for manufacturing processes.

38. Which of the following statements describes a benefit of a spaghetti diagram?

    a. It shows the shape of a data distribution.

    b. It makes inefficiencies in layout and process flow evident.

c. It identifies the root causes of a problem.

d. It identifies the variability inherent in a process.

39. What is the Japanese term in lean that means breakthrough improvement or big change?

a. Kaizen event

b. Kaizen blitz

c. Kaikaku

d. Kanban

40. The Six Sigma team collaborates with designers to ensure that final design outputs include which of the desired attributes?

a. Design for cost, design for test, and design for maintainability

b. Design for ability, design for cost, and risk reduction

c. Design for repeatability, design for waste prevention, and design for reproducibility

d. Design for recyclability, design for gage repeatability and reproducibility (GR&R), and design for feasibility

41. An onsite verification activity to ensure compliance with requirements is known as a(n):

a. functional inspection.

b. first-article inspection.

c. 100 percent inspection.

d. audit.

42. Verification can be defined as:

a. confirmation, through the provision of objective evidence, that specified requirements have been fulfilled.

b. developing certain fixed points during the process.

c. a partial cross-function team review of the project prior to full completion.

d. having the champion engage the team and senior managers to approve evidence that they have met key deliverables.

43. Validation can be defined as:
    a. evaluating a process or product and determining what might cause it to fail.
    b. making educated assessments of potential failures.
    c. confirmation, via objective evidence, that the requirements for a specific intended use or application have been fulfilled.
    d. proof of concept at the final design stage that confirms the initiative.
44. The primary purpose of a failure mode and effects analysis is the study of:
    a. robustness.
    b. prevention.
    c. risk.
    d. opportunity.
45. The fundamental reason for using an FMEA is to:
    a. schedule the milestones of a project.
    b. utilize spreadsheet software programs.
    c. use design experts in the process of FMEA.
    d. predict the highest risk of things that could go wrong.
46. What is the typical action taken when a high-risk priority number (RPN) is identified during a process FMEA?
    a. Ignoring the risk due to its complexity
    b. Implementing corrective actions to reduce the risk
    c. Delegating responsibility to another department
    d. Reducing the scope of the process to avoid the risk
47. A line item in an FMEA has a severity of 7, an occurrence score of 5, and a detection score of 3. What is the RPN?
    a. 12
    b. 15
    c. 35
    d. 105

Use the following FMEA to answer questions 48–50.

	Current Process				Action Results			
Row	Severity	Occurrence	Detection	RPN	Severity	Occurrence	Detection	RPN
1	3	5	10	150	3	5	4	60
2	5	10	3	150	3	5	3	45
3	10	3	5	150	10	3	1	30

48. On which row of the FMEA should a process improvement team focus their efforts for the current state of the process?

 a. 1

 b. 2

 c. 3

 d. The RPNs are the same; no special effort is necessary.

49. Give the rows of the current state of the process in order of highest to lowest risk.

 a. 2, 3, 1

 b. 3, 1, 2

 c. 3, 2, 1

 d. The RPNs are the same, so no ranking is needed.

50. After action has been taken, which row now shows the highest risk?

 a. Row 1

 b. Row 2

 c. Row 3

 d. None; the risk has been mitigated.

51. The focus of a design FMEA is:

 a. prototype brainstorm.

 b. product design.

 c. product packaging.

 d. design engineer training.

SOLUTIONS

1. b; The true value of Six Sigma is the bottom-line results gained by an organization. [I.A.1]

2. a; The threads common to the Six Sigma philosophy are found in Chapter 1 of *The ASQ Certified Six Sigma Green Belt Handbook*, with the best choice of this group being to reduce variation. The other three are true and can be counted as subsets or results of reducing variation in a process, as Deming and Juran have suggested in their writings. [I.A.1]

3. d; The lack of management commitment and involvement has led to a very large failure rate for Six Sigma implementation. The use of Deming's Fourteen Points, the Management's Seven Deadly Diseases, or Juran's Trilogy could help in this regard. [I.A.1]

4. d; Shewhart created the control chart in the 1920s, and Bell Labs coined the term *statistical quality control*. The first control chart developed by Shewhart was the P Chart. [I.A.1]

5. c; Also called the Ishikawa diagram. Note: Deming tried to Americanize this tool by calling it a feather diagram, but the term never stuck. Taguchi is most known for the *loss function* and design of experiments (DOEs) using orthogonal arrays. [I.A.1]

6. b; Juran wrote a great deal on issues of management over his long career. Note that he was introduced to this by watching the Hawthorne experiments as they developed at the Western Electric plant in Cicero, IL, when he was a young quality engineer. [I.A.1]

7. d; The inputs-process-output (IPO) model is the starting point of a flow diagram that will define a process. *Plan-do-check-act* (PDCA) is used more as a cycle analysis; however, the International Organization for Standardization (ISO) community recognizes this in the management system standards (MSS) as a form of a process. Both *a* and *b* are tools used in Six Sigma but are not process-approach tools. PDCA is a problem-solving tool. The IPO is a process-approach tool. [I.A.1]

8. c; The *A* in SMART goals stands for *achievable*. Given available resources and constraints, the goal should be realistic and feasible. [I.A.2]

9. d; Project management has been using SMART goals for many years. [I.A.2]

10. a; The use of SMART goals is to provide objectives for the team to meet when working through projects. [I.A.2]

11. c; Remember that a key driver of Six Sigma is to reduce variation in the process, so KPI for Six Sigma practitioners is based on the *key process indicators.* [I.A.3]

12. a; Remember that we are looking at processes, so ensuring quantifiable measurement is the key to answering this question. Both *b* and *c* are distractors that come from the SMART goal concepts. [I.A.3]

13. b; The management team or project champions should be setting the boundaries for the Six Sigma team, even when there are Master Black Belts involved. Setting boundaries also helps to prevent "scope creep." [I.A.3]

14. c; One of the most crucial considerations when selecting KPIs is to align them with the organization's strategic objectives, ensuring that they reflect what the organization wants to achieve. [I.A.3]

15. c. The balanced scorecard track measures from the financial, employee learning and growth, customer, and internal process perspectives. [I.A.3]

16. d; The balanced scorecard provides an approach to measuring multiple aspects of a business, including the financial aspect. [I.A.3]

17. c; This question may be difficult for some; remember to focus on value and waste elimination proactively. Lean works to eliminate muda anywhere it is found. [I.B.1]

18. a; This should be an easy answer as the other options are separate tools and not directly associated with the Toyota Production System House. [I.B.1]

19. a; Of the four possibilities here, only *a* drives to remove waste, which is one of the keys of lean. [I.B.1]

20. a; Typically, working to accelerate value-added functions has been the drive for managers who direct the teams to work on those areas. [I.B.1]

21. d; Note that even with the drive to improve the efficiency of operations, most attempts at sustaining 5S do not work well for organizations. [I.B.2]

22. b; In the 5S sort step, unneeded items are removed from the area. [I.B.1]

23. c; In the 5S shine step, the work area and equipment are cleaned. [I.B.2]

24. a; In the 5S standardize step, the team creates checklists, standards, and work instructions to keep the work area clean and orderly. [I.B.2]

25. b; In the 5S sustain step, management gives its support, and employees are empowered to maintain the gains achieved. [I.B.1]

26. b; The theory of constraints (TOC) is a problem-solving tool that focuses on the bottlenecks/weakest links in a process. [I.B.1]

27. c; While reduced waste, customer satisfaction, and one-piece flow are intermediate goals, the end goal of lean is perfection. [I.B.1]

28. c; A value stream is a series of activities that an organization performs, such as ordering, designing, producing, and delivering products and services. *Value stream maps* are constructed for the as-is (current), could-be, and should-be states. [I.B.2]

29. b; The symbol below denotes a production kanban in a value stream map. [I.B.2]

30. a; Inspections are a form of overprocessing waste classified as non-value added activities on a value stream map. [I.B.2]

31. b; Unused space, defects, and rework are all examples of waste. [I.B.2]

32. a; Takt is a German word meaning to set a pace. It is a method that sets the pace of industrial manufacturing lines so that production cycle times can be matched to customer demand rates. [I.B.1]

33. b; Takt time is the rate in time per unit at which the process must complete units to meet customer demand. It is calculated as the ratio of the total time available divided by the number of units the customer requires. Here, we have eight hours of time, which equals 480 minutes, and a customer demand of 320 units per day. [I.B.1]

$$\textit{Takt time} = \frac{\textit{Time available}}{\textit{Units required}} = \frac{480 \text{ min}}{320 \text{ units}} = 1.5 \text{ min / unit}$$

34. d; Takt time analysis can be used during the improvement phase to evaluate workflow. [I.B.1]

35. d; Just-in-Time (JIT) is an optimal material requirement planning system for a manufacturing process. Holding place and optimal materials are counter to lean methodologies and are not correct here. [I.B.1]

36. b; The two pillars of the Toyota Production System (TPS) are just-in-time and jidoka, or *autonomation.* [I.B.1]

37. b; A spaghetti diagram is best described as a graphic tool intended to show the flow or path that people, products, or paperwork take during the course of normal operations. Some might argue *d* is the correct answer, but there may not be multiple people or processes involved. [I.B.1]

38. b; A spaghetti diagram makes layout and process flow inefficiencies evident. [I.B.1]

39. c; Kaikaku is a breakthrough improvement. Kaizen refers to small event changes, and both events and blitzes are used for small changes. [I.B.1]

40. a; Cost, test, and maintainability would be the three primary areas the design engineers need to include in their plans for the new product. Some other elements of the distractors may come into play as we move to sustainability, but *a* is the only fully correct answer here. [I.C.1]

41. d; An onsite verification activity to ensure compliance with requirements is known as an audit. [I.C.1]

42. a; Verification is the confirmation, through objective evidence, that specified requirements have been met. [I.C.1]

43. c; Validation is the confirmation, through objective evidence, that the requirements for a specific intended use or application have been met. [I.C.1]

44. c; FMEA is primarily a risk tool to ensure customer satisfaction. Other risk tools are the strengths-weaknesses-opportunities-threats (SWOT) and PESTLE. [I.C.2]

45. d; FMEA is used to predict states with the highest risk based on severity, probability of occurrence, and probability of detection. Common forms are the design FMEA, process FMEA, system FMEA, or machinery FMEA. [I.C.2]

46. b; When a high RPN is identified, the primary action is implementing corrective actions to reduce the risk associated with the failure mode. [I.C.2]

47. d; $\text{RPN} = S \times O \times D = 7 \times 5 \times 3 = 105$. [I.C.2]

48. c; If there is a tie in RPNs, the team should compare the severity ratings and act on the row with the highest severity score. Here, Row 3 shows a severity of 10. This should be the team's focus. [I.C.2]

49. c; Since the RPNs are tied, we rank the risks of the rows by severity. Row 3 has the highest severity, followed by Row 2, then Row 1. [I.C.2]

50. a; The RPN of Row 1 is now the highest risk with $\text{RPN} = 60$. [I.C.2]

51. b; The design FMEA (DFMEA) looks at a product design. [I.C.3]

Section II
Define Phase

(Exam Targets 20 Questions)

A. Project identification

1. Project selection

Describe the project selection process and what factors should be considered in deciding whether to use the six sigma DMAIC methodology or another problem-solving process. (Understand)

2. Process elements

Define and describe process components and boundaries. Recognize how processes cross various functional areas and the challenges that result for process improvement efforts. (Analyze)

3. Benchmarking

Understand various types of benchmarking, including competitive, collaborative, and best practices. (Understand)

4. Process inputs and outputs

Identify process input and output variables and evaluate their relationships using the supplier, inputs, process, output, customer (SIPOC) model. (Analyze)

5. Owners and stakeholders

Identify the process owners and other stakeholders in a project. (Apply)

B. Voice of the customer (VOC)

1. Customer identification

Identify the internal and external customers of a project, and what effect the project will have on them. (Apply)

2. Customer data

Collect feedback from customers using surveys, focus groups, interviews, and various forms of observation. Identify the key elements that make these tools effective. Review data collection questions to eliminate vagueness, ambiguity, and any unintended bias. (Apply)

3. Customer requirements

Use quality function deployment (QFD), Critical to X (CTX when 'X' can be quality, cost, safety, etc.), Critical to Quality tree (CTQ), and Kano model to translate customer requirements statements into product features, performance measures, or opportunities for improvement. Use weighting methods as needed to amplify the importance and urgency of different kinds of input; telephone call vs. survey response; product complaint vs. expedited service request. (Apply)

C. Project management basics

1. Project methodology

Define and apply agile and top-down project management methods. (Apply)

2. Project charter

Define and describe elements of a project charter and develop a problem statement that includes baseline data or current status to be improved and the project's goals. (Apply)

3. Project scope

Help define the scope of the project using process maps, Pareto charts, and other quality tools. (Apply)

4. Project metrics

Help develop primary metrics (reduce defect levels by *x* amount) and consequential metrics (the negative effects that making the planned improvement might cause). (Apply)

5. Project planning tools

Use work breakdown structures (WBS), Gantt charts, critical path method (CPM), program evaluation and review technique (PERT) charts, and tollgate reviews to plan projects and monitor their progress. (Apply)

6. Project documentation

Describe the types of data and input needed to document a project. Identify and help develop appropriate presentation tools (storyboards, spreadsheet summary of results, etc.) for phase reviews and management updates. (Apply)

7. Project risk analysis and management

Describe the elements of project risk analysis, including feasibility, potential impact, RPN, and risk management. Identify the potential effect risk can have on project goals and schedule, resources (materials and personnel), business continuity planning, costs and other financial measures, and stakeholders. (Understand)

8. Project closure

Review with team members and sponsors the project objectives achieved in relation to the charter and ensure that documentation is completed and stored appropriately. Identify lessons learned and inform other parts of the organization about opportunities for improvement. (Apply)

D. Management and planning tools

Define, select, and apply these tools: 1) affinity diagrams, 2) interrelationship digraphs, 3) tree diagrams, 4) prioritization matrices, 5) matrix diagrams, 6) process decision program charts (PDPCs), 7) activity network diagrams, and 8) SWOT analysis. (Apply)

E. Business results for projects

1. Process performance

Calculate process performance metrics such as defects per unit (DPU), rolled throughput yield (RTY), cost of poor quality (COPQ), defects per million opportunities (DPMO), sigma levels, and process capability indices. Track process performance measures to drive project decisions. (Analyze)

2. Communication

Define and describe communication techniques used in organizations: top-down, bottom-up, and horizontal. (Apply)

F. Team dynamics and performance

1. Team stages and dynamics

Define and describe the stages of team evolution, including forming, storming, norming, performing, adjourning, and recognition. Identify and help resolve negative dynamics such as overbearing, dominant, or reluctant participants, the unquestioned acceptance of opinions as facts, groupthink, feuding, floundering, the rush to accomplishment, attribution, discounts, digressions, and tangents. (Understand)

2. Team roles and responsibilities

Use tools, such as RACI, to describe and define the roles and responsibilities of participants on six sigma and other teams, including Black Belt, Master Black Belt, Green Belt, Champion, Executive, coach, facilitator, team member, sponsor, and process owner. (Apply)

3. Team tools and decision-making concepts

Define and apply team tools such as brainstorming and decision-making concepts such as nominal group technique, and multi-voting. (Apply)

4. Team communication

Identify and use appropriate communication methods (both within the team and from the team to various stakeholders) to report progress, conduct reviews, and support the overall success of the project. (Apply)

QUESTIONS

1. The define phase of the DMAIC model serves two primary purposes, which are to define the project management process and the:

   a. problem or issue to be worked on.

   b. desired project outcome.

   c. budgetary constraints.

   d. solution to the problem.

2. The project selection group (Master Black Belts, Black Belts, Key Executives, and Champions) should:
   a. be certain to include other critical organizational goals.
   b. identify the cost reduction terms first.
   c. always select the projects with known solutions.
   d. establish a set of criteria for project selection.
3. When planning to study a process or system, it is very important to first identify the process:
   a. environmental challenges.
   b. boundaries.
   c. history.
   d. business plan.
4. Recognizing that if one process factor is changed or influenced in some particular way, then something different might happen downstream in the process is commonly referred to as:
   a. an effective business plan.
   b. part of a good strategic plan.
   c. critical analysis.
   d. systems thinking.
5. Going to a local fast-food restaurant to understand how its processes work would be an example of which type of benchmarking?
   a. Internal
   b. Competitive
   c. Functional
   d. Collaborative
6. The following are types of benchmarking:
   a. internal, external, and subgroup.
   b. team, individual, competitive, and extreme.
   c. internal, competitive, functional, and collaborative.
   d. external, community, and cultural.

7. Which tool helps a team to quickly familiarize themselves with the process at an organizational level and visually understand the scope of the project?
   a. SIPOC chart
   b. Cause-and-effect matrix
   c. Scatter diagram
   d. Critical to quality tree
8. Stakeholders (e.g., customers, suppliers, employees, investors) generally:
   a. have a vested interest in the process and/or its products and outputs.
   b. are responsible for adding more funding to the project as needed.
   c. meet with the Six Sigma team to report on the progress of the project's effectiveness.
   d. are always subject matter experts within the project's scope.
9. Those who have responsibility for the execution and implementation of specific tasks or activities being worked on are:
   a. subject matter experts.
   b. process owners.
   c. top management.
   d. stakeholders.
10. A human resources department is addressing its onboarding process to eliminate redundant steps. For this project, a new hire would be considered a(n):
    a. internal customer.
    b. external customer.
    c. process owner.
    d. champion.
11. There are several tools for gathering customer data, including:
    a. observation, cause-and-effect diagram, and Pareto charts.
    b. surveys, quality function deployment (QFD), interviews, and focus groups.
    c. phone surveys, mailed questionnaires, online surveys, and FMEAs.
    d. secret shopper, random selection X-ray scans, and SPC.

12. Statistically, what is the most valid procedure for collecting customer data?
    a. Randomly selected large groups for interviews
    b. Randomly selected large groups for surveys
    c. Randomly selected small groups for mailed questionnaires
    d. Small focus groups
13. A QFD is also known as the house of:
    a. design and redesign.
    b. quality.
    c. linkages.
    d. product planning.
14. Which graphical tool best shows the relationship between product features and customer satisfaction?
    a. PDCA
    b. Spaghetti diagram
    c. Ishikawa chart
    d. Kano model
15. The primary focus of a Kano analysis is to:
    a. determine takt time.
    b. discover customer needs and wants.
    c. improve throughput.
    d. move teams from storming to performing.
16. The traditional Kano model y-axis shows:
    a. source dysfunctional/service fully functional.
    b. delighters/satisfiers.
    c. dissatisfaction/satisfaction.
    d. one-dimensional versus two-dimensional.

17. In the Kano model, which category represents features that, when present, lead to higher customer satisfaction, but when absent, do not result in dissatisfaction?
    a. Basic needs
    b. Performance needs
    c. Delighters
    d. Indifferent needs
18. A Kano analysis shows that battery life is a one-dimensional requirement. This means that:
    a. the customer will be delighted with the product.
    b. the longer the battery life, the happier the customer will be.
    c. the customer is indifferent to the length of battery life.
    d. increasing battery life will not improve customer satisfaction.
19. Critical-to-X factors found in Six Sigma should be focused on:
    a. customers.
    b. suppliers.
    c. the management team.
    d. process.
20. The agile approach aims to:
    a. reduce cycle time and waste.
    b. identify customer wants and needs.
    c. map processes using symbols.
    d. increase the scope of projects.
21. The term coined in the lean approach/lean thinking that focuses on reducing cycle time and waste is called the:
    a. Ishikawa diagram.
    b. Kano model.
    c. loss function.
    d. agile approach.

22. Effective project planning requires skills in the following areas:

    a. information processing, communication, resource negotiations, securing commitments, incremental planning, assuring measurable milestones, and facilitating top management involvement.

    b. communication, statistical process control, risk-based thinking, problem-solving, resource negotiations, facilitation skills, and team negotiating skills.

    c. information processing, decision-making, legal analysis, and securing commitments.

    d. information processing, advance product quality planning, risk-based thinking, covert processes, influential interventions, and facilitating top management involvement.

23. A project charter should contain which elements?

    a. Problem statement, purpose, cost projection, scope, and next steps

    b. Scope, team member names, cost projection, benefits, and results

    c. Vision statement, purpose, strategic plan, benefits, and results

    d. Problem statement, purpose, benefits, scope, and resources needed

24. A tool that can help refine the scope of a project is a(n):

    a. process decision program chart.

    b. affinity diagram.

    c. activity network diagram.

    d. Pareto chart.

25. A key aspect of project metrics is to start:

    a. by tracking only what management believes is important.

    b. tracking the data as early as possible to benchmark the current state.

    c. at the first milestone in order to get meaningful data.

    d. by selecting easy-to-collect measurements.

26. A project planning tool that shows the early start, early finish, late start, and late finish times for each activity is known as a(n):

    a. Gantt chart.

    b. affinity diagram.

c. PERT chart.

d. work breakdown.

27. The following graphic is an example of a(n):

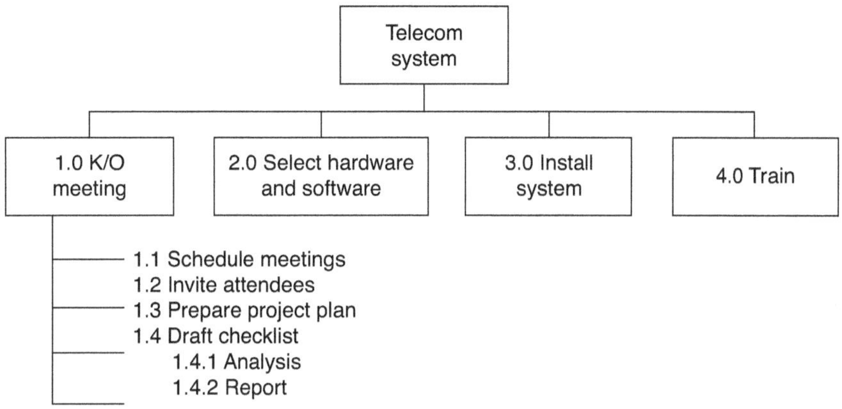

a. activity network diagram.

b. affinity diagram.

c. force field analysis.

d. work breakdown structure.

28. A Green Belt creates a PERT chart for a project. Activities B, C, and F are on the critical path. What is the result if Activity B is delayed by 3 days?

a. The entire project will be delayed by 3 days.

b. Activity B will no longer be on the critical path.

c. The project will not be delayed if Activity A is completed early.

d. Activity C will start 6 days late.

29. What can be used as an input to a Gantt chart?

a. SIPOC diagram

b. Work breakdown structure

c. Project goal statement

d. Affinity diagram

30. Which document may be used in a tollgate meeting for the measure phase of a Six Sigma project?

a. Root cause analysis results

b. A process map

c. Brainstormed solutions

d. A control plan

31. Which document would be most appropriate to present in a tollgate meeting in the analyze phase of a Six Sigma project?

a. Root cause analysis

b. Control plan

c. Pilot test results

d. Process map

32. Which of the following tools has been developed to ensure that the project is on the right path to ensure the desired outcomes?

a. Force field analysis

b. Tollgates

c. Check sheets

d. Plan-do-check-act

33. Successful risk analysis depends, in part, on ensuring that:

a. all internal and external customers contribute safety data prior to the first risk analysis meeting.

b. warranty and reliability data are treated as the most important inputs.

c. all appropriate organizations are represented during risk analysis meetings, with all aspects of potential risk considered.

d. risk mitigation and verification are performed only at the final stages of the analysis.

34. Which scenario is business continuity designed to deal with?

a. Monitoring normal supply chain issues for day-to-day operations

b. Ensuring that operations can continue in adverse conditions

c. Dealing with a design complaint from a key customer during the launch of a product

d. Finding failure modes for customer complaints

35. Risk management is defined in the ISO 31000:2018 as:

   a. the natural boundaries of a process within specified confidence levels.

   b. a systematized group of activities to recognize and evaluate the potential failure of a product or process.

   c. a characteristic of a product or service that is essential to ensure customer satisfaction.

   d. coordinated activities to direct and control an organization concerning risk.

36. Risk management can be defined using the PDCA process. Which item is part of *Act*?

   a. Implementing a test of the chosen risk management technique(s)

   b. Monitoring the implemented techniques for adequacy of protection

   c. Assessing the effectiveness of the technique and making necessary adjustments

   d. Identifying and defining its potential exposures to loss

37. The final step that is performed to prove that the project met established goals and objectives and that all documentation was completed and stored is called:

   a. the charter grand finale.

   b. an independent third-party review.

   c. Six Sigma team project termination.

   d. the project closure.

38. One activity of project closure is to:

   a. identify and share lessons learned with the organization.

   b. justify the cost of the project.

   c. develop a control plan and return the process to the process owner.

   d. design for Six Sigma.

39. Which tool graphically shows the interdependence between tasks while highlighting key tasks, time to accomplishment, flow paths, and so on?

   a. Activity network diagram

   b. Cause-and-effect diagram

c. Pareto chart

d. Control chart

40. What important planning tool is founded on the premise that solid planning helps prevent undesirable surprises and saves valuable resources?

 a. Early executive adventure

 b. Advance quality planning

 c. Association matrix

 d. The seven-step improvement model

41. Which tool is used to produce many possible answers to an open-ended question?

 a. Venn diagram

 b. Fishbone diagram

 c. Affinity diagram

 d. Cause-and-effect diagram

42. A SWOT analysis consists of:

 a. satisfaction, work, output, and target.

 b. strength, weakness, opportunity, and threat.

 c. solution, work, overall, and test.

 d. severity, worst, operating, and time.

43. The first thing to know when planning an audit is:

 a. how many internal and external audits have previously been conducted.

 b. whether all the important rules are in writing.

 c. whether the employees have been coached in advance.

 d. what the standards or criteria are to be audited against.

44. Benchmarking is the process of looking at one system(s) and applying the concepts observed to:

 a. the lessons learned fact sheet.

 b. another system.

 c. all prototype production.

 d. the deployment of first production.

45. A flowchart is a picture of the separate steps of a process in sequential order, including:

   a. activities and diagrams.

   b. materials, services, people, decisions, time, and process measurements.

   c. activities, stakeholders, strategic plan, and risk analysis.

   d. aspects, impacts, hazardous analysis, safety concerns, and environmental risks.

46. Once focus group planning is completed, the focus groups can be:

   a. developed, exercised, data collected, and data analyzed.

   b. identified, conducted, data collected, and data analyzed.

   c. identified, trained, interviewed, and surveys analyzed.

   d. identified, confirmed, coached, and data collected.

47. What is the main purpose of using process decision program charts (PDPCs) in Six Sigma projects?

   a. To create flowcharts of the entire project process

   b. To visualize the relationships between different process steps

   c. To identify potential risks and develop contingency plans

   d. To analyze process variations using statistical techniques

48. A strategic technique used to assess an organization's competitive position is:

   a. SWOT.

   b. FMEA.

   c. PESTLE.

   d. RACI.

49. The four elements of the Cost of Quality are:

   a. Internal, external, supplier, warranty

   b. Internal, external, appraisal, rework

   c. Internal, external, customer, design

   d. Internal, external, appraisal, prevention

50. Warranty costs are an example of which Cost of Quality?
    a. Internal
    b. External
    c. Appraisal
    d. Prevention
51. Which index is used to identify the total number of defects/deficiencies that have occurred and standardized to a common total number of possibilities?
    a. DPO
    b. DPU
    c. DPMO
    d. RTY
52. Which index is used to identify the first pass rate through a series of processes?
    a. DPO
    b. DPU
    c. DPMO
    d. RTY
53. A ring gauge is used to check the diameter of a metal shaft in three places. An inspector tests 1000 shafts and finds six defects. What is the DPMO?
    a. 0.002
    b. 0.006
    c. 2000
    d. 6000
54. Calculate the rolled throughput yield of a drilling process with three consecutive machine steps that have yields of 98%, 96%, and 96%, respectively.
    a. 2.9%
    b. 0.97%
    c. 90%
    d. 97%

55. The defect level of a 6σ-level process, after a 1.5σ shift, is equal to:

    a. 2 parts per billion.

    b. 3.4 parts per million.

    c. 34 parts per million.

    d. 2700 parts per million.

56. What is a key advantage of bottom-up communication in organizations?

    a. Efficient dissemination of top-level decisions

    b. Quick decision-making at the senior management level

    c. Enhanced employee engagement and empowerment

    d. Facilitation of direct communication between peers

57. Flow utilized when top management, or the executive sponsor of the project, is providing instruction, communicating policies, or providing feedback on project performance is called:

    a. lateral flow.

    b. downward flow.

    c. diagonal flow.

    d. matrix flow.

58. What is the primary purpose of horizontal communication in organizations?

    a. To convey performance feedback from supervisors to employees

    b. To share information and collaborate among peers or colleagues

    c. To deliver strategic directives from top management

    d. To provide opportunities for employees to voice concerns to management

59. Teams typically go through the following sequence or stages of evolution:

    a. Forming, challenging, testing, and agreeing.

    b. Forming, storming, norming, and performing.

    c. Circling, advancing, organizing, and accomplishing.

    d. Honeymooning, presenting, advocating, and completing.

60. A function of team storming includes:
    a. team members getting to know each other.
    b. team members voicing their opinions.
    c. team members resolving their conflicts.
    d. team members realizing interdependence.
61. Negative team dynamics are often reflective of:
    a. negative issues within the organizational culture.
    b. a lack of training in how to be a team member.
    c. a team leader who refused to lead the team.
    d. team members with too many distractions and a lack of commitment.
62. One way to reduce groupthink in a team is to:
    a. enforce discipline.
    b. encourage mutual respect.
    c. use the devil's advocate approach.
    d. bring in independent, outside members to the meeting.
63. One factor that has helped Six Sigma be successful is:
    a. the amount of feedback built into the system.
    b. the depth of historical background attributed to it.
    c. the responsibility of business unit leaders for profit and loss.
    d. the structure it demands of the organization.
64. An individual trained in Six Sigma methodologies, statistical tools, basic financial tools, change management, risk assessment, and project management is a:
    a. process owner.
    b. Project Champion.
    c. Black Belt.
    d. Green Belt.

65. The Master Black Belt is usually associated with an individual who typically:

    a. retains their regular position but is trained in the tools, methods, and skills necessary to conduct Six Sigma improvement projects.

    b. is assigned full-time to train, mentor, and lead improvement projects.

    c. is assigned full-time to train, mentor, and lead the strategy for chartering the organization's improvement projects.

    d. has the authority or ability to make changes in the process as required.

66. RACI commonly stands for:

    a. registration, accreditation, conformance, and independence.

    b. responsible, accountable, consulted, and informed.

    c. reliable, act, check, and improve.

    d. reliability, availability, consistency, and improvement.

67. The added role found in RASCI, which is a sister to the RACI matrix, is for the role of:

    a. support.

    b. staff.

    c. senior SME.

    d. safety.

68. In which role in the RACI model must a decision be discussed with the individual before a final decision is made?

    a. Responsible

    b. Accountable

    c. Consulted

    d. Informed

69. The tool used in typical Green Belt projects to help define the roles of team members is:

    a. PFMEA.

    b. SWOT.

    c. PESTLE.

    d. RACI.

70. The technique in which team members consolidate all previously generated ideas, number or letter them for identification, and quickly prioritize the ideas is known as:

    a. multi-voting.

    b. affordability consolidation.

    c. assigned severity.

    d. RPN assessment.

71. The table shown here is an example of:

	Member 1	Member 2	Member 3	Member 4	Member 5	Member 6	Member 7	Member 8	Member 9	Total
A	5	3	4	5	5	3	4	5	5	39
B	2	4	3	3	4	4	3	3	4	30
C	1	5	5	4	3	5	5	4	3	35
D	3	3	1	2	2	3	1	2	2	19
E	4	1	2	1	1	1	2	1	1	14

    a. nominal group technique (NGT).

    b. multi-voting.

    c. PESTLE.

    d. kanban matrix.

72. A convergence tool used to reduce brainstormed ideas down to the causes most likely to impact the effect is called:

    a. nominal group technique.

    b. a cause-and-effect diagram.

    c. a Pareto chart.

    d. an affinity diagram.

73. Nominal group technique is especially effective when:

    a. the team must make a decision quickly.

    b. the issue being discussed is sensitive.

    c. dominant team members are absent.

    d. all team members have the same job titles.

74. The approach that allows a team to work through a process to come to a conclusion regarding the topic at hand is called:

    a. fact-based decision-making.

    b. top-down decision-making.

    c. consensus decision-making.

    d. brainstorming.

75. One way a team can communicate its project results to the organization is to:

    a. share lessons learned.

    b. incorporate the results in the annual report.

    c. hold a focus group.

    d. benchmark against a competitor.

SOLUTIONS

1. a; In the define phase of a Six Sigma project, the project management process and the problem or issue to be worked on are defined. [II.A.1]
2. d; Establishing a set of criteria for project selection is the standard for selecting Six Sigma projects. [II.A.1]
3. b; The first step in any process work is to identify the boundaries for the process that you intend to investigate. Then go to Gemba—go and see what is actually being done. [II.A.2]
4. d; Systems thinking is defined as recognizing that if one process factor is changed or influenced in some particular way, then something different might happen downstream in the process. [II.A.2]
5. c; Understanding how processes work is an example of functional benchmarking. [II.A.3]
6. c; Common types of benchmarking include internal, competitive, functional, and collaborative. [II.A.3]
7. a; A SIPOC chart lists suppliers, inputs, process steps, outputs, and customers at a high level. It helps the team understand the process at an organizational level, as well as the scope of the project. [II.A.4]
8. a; Another term from ISO MSS is interested parties, with stakeholder being one of these. Thus, stakeholders are interested in the outcomes of the Six Sigma team. [II.A.5]
9. b; Process owners have responsibility for the execution and implementation of specific tasks or activities being worked on. [II.A.5]
10. b; The human resource employees would be internal process customers, and the new hires would be considered external customers of the process. [II.B.1]
11. b; Surveys, quality function deployment (QFD), interviews, and focus groups are all tools for gathering voice of the customer data. [II.B.2]
12. a; Remember that reasonably large groups of customers are needed to obtain complete and accurate data on each topic selected; surveys cannot do this. To improve accuracy, increase the sample size. [II.B.2]
13. b; QFD or quality function deployment is also known as the house of quality. [II.B.3]
14. d; A Kano model is a graphical tool for showing the relationship between product features and customer satisfaction. [II.B.3]

15. b; The primary focus of Kano analysis is to discover customer needs and wants. [II.B.3]

16. c; In the Kano model representation, the y axis shows customer dissatisfaction/satisfaction, as shown below.

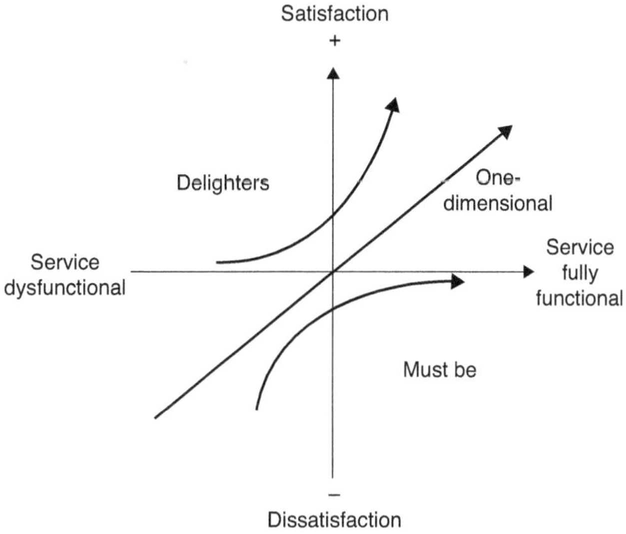

[II.B.3]

17. c; Delighters are features that, when present, exceed customer expectations and lead to higher satisfaction levels, but their absence doesn't result in dissatisfaction. [II.B.3]

18. b; Customer satisfaction increases as a product or service delivers more of a one-dimensional requirement. [II.B.3]

19. a; Customer, always. There are several critical-to-X factors, and all of them will end up with a focus on customers of the products, processes, or services. [II.B.3]

20. a; The aim of the agile approach is to reduce cycle time and waste. [II.C.1]

21. d; The agile approach focuses on reducing cycle time and waste. [II.C.1]

22. a; Many skills are needed throughout the entire project. This question focuses on the startup of the project where information processing, communication, resource negotiations, securing commitments, incremental planning, assuring measurable milestones, and facilitating top management involvement are important. [II.C.1]

23. d; Some charters may include cost numbers, but this is not a requirement. Focus on getting management and champion buy-in. Also note that some organizations use the A3 as their charters, which should have the same information on one page. [II.C.2]

24. d; A Pareto chart can help set the scope of a project by revealing the largest contributors to a problem. [II.C.3]

25. b; We should always aim to get as much data as possible prior to the official launch of a project. The Hawthorne Effect will skew the data if people realize they are being watched. [II.C.4]

26. c; A PERT chart shows the early & late start and early & late finish times for each activity in the project plan. [II.C.5]

27. d.; The figure below shows an example of a work breakdown structure.

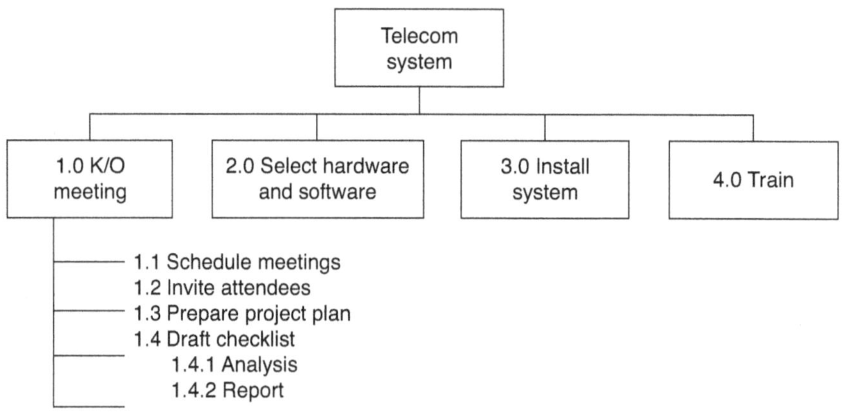

[II.C.5]

28. a; Any delay in an activity on the critical path will result in a project delay. [II.C.5]

29. b; The work breakdown structure is developed to assign and track work. The resulting tasks can be used as inputs for a Gantt chart. [II.C.5]

30. b; A process map is a key deliverable for the measure phase of a Six Sigma project. [II.C.6]

31. a; Root cause analysis is a key deliverable for the analyze phase of a Six Sigma project. [II.C.6]

32. b; Tollgates, also known as advance product quality planning (APQP), are used to verify the project process. [II.C.6]

33. c; Successful risk analysis depends, in part, on having all appropriate organizations represented in the risk analysis meetings. As many interested parties as possible need to be engaged in the process. This includes internal and external customers. [II.C.7]

34. b; Business continuity is a design for when things do not go as planned (for instance, like what happened during the worldwide pandemic). It is designed to ensure that operations can continue in adverse conditions. [II.C.7]

35. d; There is a series of ISO risk documents in the ISO 31000 series. ISO 31010 lists a number of risk registers that can be very useful during Six Sigma projects. [II.C.7]

36. b; In the act phase, we monitor the implemented techniques for adequacy of protection. [II.C.7]

37. d; The project closure could include things like an A3 and other pertinent information from the project to demonstrate that goals have been met. [II.C.8]

38. a; When closing the project, the team should identify lessons learned and share them with the rest of the organization. [II.C.8]

39. a; An activity network diagram shows the interdependence between tasks while highlighting key tasks, time to accomplishment, flow paths, and so on. [II.D]

40. b; The AQP was renamed APQP for the automotive industry to highlight that the design/product engineers need to be involved during the initial planning to help prevent surprises. [II.D]

41. c; The affinity diagram is used to produce and categorize as many possible answers to an open-ended question. [II.D]

42. b; A SWOT analysis consists of identifying the internal strengths and weaknesses of an organization, as well as outside opportunities and threats. [II.D]

43. d; The first step in planning an audit is to identify the desired requirements of the standard to be reviewed. A good reference here is ISO 19011 for conducting audits. [II.D]

44. b; Benchmarking is the process of looking at one system and applying the concepts observed to another system. Getting people to observe one thing and then apply it in another area, which could be different from what was initially observed, is often a challenge. [II.D]

45. b; A flowchart is a display of the separate steps in a process shown in sequential order, including materials, services, people, decisions, time, and process measurements. [II.D]

46. b; Once the focus group planning is complete, the next steps are to identify participants, conduct the focus group, and collect and analyze the resulting data. [II.D]

47. c; PDPC is used to systematically identify potential risks in a process, assess their impact, and develop contingency plans to address them. [II.D]

48. a; SWOT lists the strengths, weaknesses, opportunities, and threats of an organization to assess its competitive position. [II.D]

49. d; The Cost of Quality comprises four costs: internal, external, appraisal, and prevention. [II.E.1]

50. b; Warranty costs are an example of an external quality cost. The external costs of quality are those defects that the customer sees. [II.E.1]

51. c; Defects per million opportunities are used to identify the total number of defects/deficiencies that have occurred, standardized to a common total number of possibilities. [II.E.1]

52. d; Rolled throughput yield is used to identify the first pass rate through a series of processes. [II.E.1]

53. c; There are three opportunities for defects on each shaft. The defects per million opportunities are calculated using:

$$DPMO = \frac{Defects}{Opportunities} \times 10^6 = \frac{6}{3 \times 1000} \times 10^6 = 2000.$$

[II.E.1]

54. c; To calculate rolled throughput yield, we multiply the individual machine yields:

$$RTY = 0.98 \times 0.96 \times 0.96 = 0.903 \times 100\% \approx 90\%.$$

[II.E.1]

55. b; For a normally distributed process output, a 6σ-level process, after a 1.5σ shift in the mean, results in a defect level of 3.4 parts per million. [II.E.1]

56. c; Bottom-up communication enhances employee engagement and empowerment by allowing employees to share their ideas, concerns, and feedback directly with higher-level management. [II.E.2]

57. b; Downward flow, also known as top-down flow, occurs when top management, or the project's executive sponsor, provides instruction, communicates policies, or provides feedback on project performance. [II.E.2]

58. b; Horizontal communication serves to share information, collaborate, and exchange ideas among peers or colleagues at the same organizational level. [II.E.2]

59. b; Teams typically go through the forming, storming, norming, and performing stages of evolution. This has been shown time and time again and has been well documented in many references. Also, remember that as new team members join the group, the process may need to start over. [II.F.1]

60. b; Team members voicing their opinions, and often disagreeing, is an example of team storming behavior. [II.F.1]

61. a; Negative team dynamics are often reflective of negative issues within the organizational culture. [II.F.1]

62. d; A countermeasure to groupthink, in which there is no public disagreement, is to bring in independent, outside members to the team meeting. [II.F.1]

63. d; The structure that Six Sigma demands on the organization has helped it be successful. This will be true as long as the management team is engaged and active. Remember that Deming identified management as 85 percent of the problem in many organizations, and this is where part of the discipline is needed. [II.F.1]

64. c; A Black Belt is trained in Six Sigma methodologies, statistical tools, basic financial tools, change management, risk assessment, and project management. [II.F.2]

65. c; A Master Black Belt is a Six Sigma role associated with an individual typically assigned to train, mentor, and lead the strategy for chartering the organization's improvement projects full-time. [II.F.2]

66. b; RACI stands for responsible, accountable, consulted, and informed. [II.F.2]

67. a; A support person can include any number of personnel, and this is the best answer here, as it covers what the team may need. [II.F.2]

68. c; The RACI model is a matrix that lists stakeholders and tasks, indicating who is responsible, accountable, consulted, and informed for each task. A stakeholder in the "consulted" role must be asked about a decision before the final decision is made. [II.F.2]

69. d; A RACI diagram is used to help define the roles of team members. [II.F.2]

70. a; When using multi-voting, team members consolidate all previously generated ideas, number or letter them for identification, and quickly prioritize the ideas. [II.F.3]

71. b; This table shows a matrix diagram that is a multi-voting tally. [II.F.3]

72. a; Nominal group technique is used to narrow brainstormed ideas to the causes most likely to impact the effect. [II.F.3]

73. b; The nominal group technique is especially effective when the issue being discussed is sensitive. [II.F.3]

74. c; In consensus decision-making, the team works together to come to a conclusion regarding the topic at hand. [II.F.3]

75. a; The project team can communicate their results by sharing lessons learned with the organization. [II.F.4]

Section III

Measure Phase

(Exam Targets 20 Questions)

A. Process analysis and documentation

Develop process maps and review written procedures, work instructions, and flowcharts to identify any gaps or areas of the process that are misaligned. (Create)

B. Probability and statistics

1. Basic probability concepts

Identify and use basic probability concepts: independent events, mutually exclusive events, multiplication rules, permutations, and combinations. (Apply)

2. Central limit theorem

Define the central limit theorem and describe its significance in relation to confidence intervals, hypothesis testing, and control charts. (Understand)

C. Statistical distributions

Define and describe various distributions as they apply to statistical process control and probability: normal, binomial, Poisson, chi-square, Student's *t*, and F. (Understand)

D. Collecting and summarizing data

1. Types of data and measurement scales

Identify and classify continuous (variables) and discrete (attributes) data. Describe and define nominal, ordinal, interval, and ratio measurement scales. (Analyze)

2. Sampling and data collection plans and methods

Define and apply various sampling methods (random and stratified) and data collection methods (check sheets and data coding). Prepare data collection plans that include gathering data and performing quality checks (e.g., minimum/maximum values, erroneous data, null values). (Apply)

3. Descriptive statistics

Define, calculate, and interpret measures of dispersion and central tendency. Develop and interpret frequency distributions and cumulative frequency distributions. (Evaluate)

4. Graphical methods

Construct and interpret diagrams and charts that are designed to communicate numerical analysis efficiently, including scatter diagrams, normal probability plots, histograms, stem-and-leaf plots, and box-and-whisker plots. (Create)

E. Measurement system analysis (MSA)

Calculate, analyze, and interpret measurement system capability using GR&R studies, measurement correlation, bias, linearity, percent agreement, and precision/tolerance (P/T). (Evaluate)

F. Process and performance capability

1. Process performance versus process specifications

Define and distinguish between natural process limits and specification limits, and calculate process performance metrics. (Evaluate)

2. Process capability studies

Define, describe, and conduct process capability studies, including identifying characteristics, specifications, and tolerances, and verifying stability and normality. (Evaluate)

3. Process capability (C_p, C_{pk}) and process performance (P_p, P_{pk}) indices

Describe the relationship between these types of indices. Define, select, and calculate process capability and process performance. Describe when C_{pm} measures can be used. Calculate the sigma level of a process. (Evaluate)

4. Short-term versus long-term capability and sigma shift

Describe the assumptions and conventions that are appropriate to use when only short-term data are used. Identify and calculate the sigma shift that occurs when long- and short-term data are compared. (Evaluate)

QUESTIONS

1. A graphic that shows a set of interrelated activities that transforms inputs into outputs is an:

   a. process map.

   b. spaghetti diagram.

   c. Pareto chart.

   d. activity network diagram.

2. What should be the first step in creating a process flowchart?
   a. Build the sequence of the process, putting everything into the appropriate order.
   b. Determine the various steps in the process through team brainstorming or walking the process.
   c. Draw the flowchart using the appropriate mapping symbols.
   d. Create the boundaries of the process that we intend to flowchart.
3. One of the differences between a work instruction and a procedure is that the work instruction identifies:
   a. why it is done (business reason, purpose).
   b. how it is done (step by step).
   c. when it is done (trigger).
   d. where it is done (location/process step).
4. What is the term for an unwritten process that must be followed consistently?
   a. Work instruction
   b. Product map
   c. FMEA
   d. Method
5. A graphical tool that shows the physical location of defects is known as a:
   a. relationship diagram.
   b. SIPOC chart.
   c. measles chart.
   d. spaghetti chart.
6. What tool helps us visualize the items charted as "vital few" and "trivial many" using the 80–20 rule?
   a. Design of experiments
   b. Affinity diagram
   c. Scatter diagram
   d. Pareto chart

7. What is the probability of rolling a 2 in a single toss of a standard six-sided die?
   a. 1/6
   b. 1/3
   c. 1/2
   d. 2/3
8. The general addition rule of probability states that P(A or B) =
   a. P(A) + P(B).
   b. P(A) + P(B) − P(A & B).
   c. P(A) × P(B | A).
   d. P(A & B) ÷ P(A).
9. When throwing a standard six-sided die, what is the probability of getting a 4 or a 6?
   a. 1/6
   b. 1/3
   c. 1/2
   d. 2/3
10. What is the formula for the probability of Event A or B occurring when the events are mutually exclusive?
    a. P(A or B) = P(A) + P(B)
    b. P(A and B) = P(A) × P(B)
    c. P(A and B) = P(A) + P(B) − $P(A \cap B)$
    d. P(A or B) = P(A) × P(B) − $P(A \cap B)$
11. There are two extrusion machines that produce plastic parts. Forty percent of the parts are produced on Machine A. The probability that a randomly chosen part is made on Machine A and is defective is 0.02. What is the probability that a part is defective given that it is made on Machine A?
    a. 0.008
    b. 0.05
    c. 0.20
    d. 0.42

A customer service representative has tabulated the most recent customer complaints. Use the frequency table below to answer questions 12–14:

Complaint Type	Frequency
Missing connectors	32
Wrong connectors	15
Directions not clear	20
Shipping damage	23
Lost shipment	10

12. What is the probability that a randomly chosen complaint concerns wrong connectors?

    a. 15/47

    b. 15/100

    c. 32/100

    d. 47/100

13. What is the probability that a complaint involves a shipping concern?

    a. 10/100

    b. 23/100

    c. 33/100

    d. 33/67

14. What is the probability that a connector complaint concerns missing connectors?

    a. 32/100

    b. 15/32

    c. 32/47

    d. 47/100

15. There are 15 inspection reports stored in a file folder. How many ways can five reports be chosen for an audit?

    a. 3

    b. 5

    c. 3003

    d. 360

16. How many ways can a group of 20 candidates be assigned to the chair, vice-chair, and secretary positions of a volunteer organization?

    a. 3

    b. 60

    c. 1140

    d. 6840

17. The distribution of averages tends to be normal, even when the distribution from which the averages are computed is not normal. This statement is a consequence of the:

    a. theory of constraints.

    b. Z distribution.

    c. Poisson distribution.

    d. central limit theorem.

18. Samples of size 36 are taken from a skewed population with a mean equal to 10.0 and a standard deviation equal to 10.0. The sample averages will follow which distribution?

    a. Exponential with $\mu = 10$, $\sigma = 10$

    b. Normal with $\mu = 10$, $\sigma = 0.278$

    c. Normal with $\mu = 10$, $\sigma = 1.667$

    d. Normal with $\mu = 10$, $\sigma = 10$

19. Since control charts like *X*-bar and *R* charts and *X*-bar and *s* charts are plotted with averages of the individual readings, the charts are fairly robust against:

    a. departures from normality.

    b. departures from special cause variation.

    c. alpha risk.

    d. the central limit theorem.

20. In a statistical study, what refers to the collection of all items or data under consideration?

    a. Parameter

    b. Statistic

c. Population

d. Sample

21. Greek letters are used to denote:

 a. sample sizes.

 b. probabilities.

 c. sample statistics.

 d. population parameters.

22. The following density function is used to model which distribution?

$$f(x)=\frac{1}{\sigma\sqrt{2\pi}}e^{-\frac{(x-\mu)^2}{2\sigma^2}}$$

 a. Normal

 b. Chi-square

 c. Binomial

 d. Poisson

23. Which distribution is the following probability mass function used to model?

$$P(X=x)=\frac{\lambda^x e^{-\lambda}}{x!}$$

 a. Normal

 b. Chi-square

 c. Binomial

 d. Poisson

24. What is the variance of the Poisson distribution?

 a. σ^2

 b. λ

 c. λ^2

 d. $e^{-\lambda}$

25. Which distribution requires two parameters to fully describe it?

    a. Normal

    b. Poisson

    c. Student's *t*

    d. Chi-square

26. Which distribution would be used to model the number of surface defects found on samples of 10 stainless steel panels?

    a. Normal

    b. Binomial

    c. Poisson

    d. Chi-square

27. Which distribution would be used to find the probability that three defective units will be found in a sample of 20 units?

    a. Normal

    b. Binomial

    c. Poisson

    d. Student's *t*

28. Which sampling distribution is symmetric with heavier tails than a normal distribution?

    a. Student's *t*

    b. Binomial

    c. Chi-square

    d. F

29. Which distribution is used to test the ratio of two variances?

    a. Normal

    b. Chi-square

    c. Student's *t*

    d. F

30. Which distribution is used to test a standard deviation against a standard?

    a. Normal

    b. Chi-square

    c. Student's *t*

    d. F

31. A sample of size six is randomly selected from a batch with a 10% defective rate. Find the probability that the sample has exactly one defective.

    a. 0.4096

    b. 0.3543

    c. 0.3281

    d. 0.2916

32. An electronic component manufacturer has an average defect rate of 0.01 per unit. What is the probability that an inspector finds one or fewer defects in a sample of 100 units?

    a. 0.0905

    b. 0.3679

    c. 0.7358

    d. 0.9953

33. A normal distribution can be standardized using which transformation?

    a. Z

    b. F

    c. t

    d. χ^2

34. Which of the following probability distributions has negative values?

    a. Poisson

    b. Student's *t*

    c. Binomial

    d. F

35. Find the area under the standard normal curve between 1.00 standard deviation and 2.50 standard deviations from the mean.

   a. 0.1455

   b. 0.1524

   c. 0.8500

   d. 0.8545

36. The weights of soup cans follow a normal distribution with a mean of 16.0 oz and a standard deviation of 0.005 oz. What percentage of cans weigh more than 16.0075 oz?

   a. 6.68%

   b. 44.04%

   c. 55.96%

   d. 93.32%

37. Find $P(-1.33 < Z < 2.75)$.

   a. 0.0918

   b. 0.9053

   c. 0.9082

   d. 0.9970

38. Find the probability that $Z \geq 0.50$.

   a. 0.3085

   b. 0.5000

   c. 0.6915

   d. 0.9915

39. Find the critical value for the chi-square distribution with 5 degrees of freedom and $\alpha = 0.05$.

   a. 0.711

   b. 9.488

   c. 11.070

   d. 12.592

40. Find the critical value for a Student's t statistic with 7 degrees of freedom and $\alpha = 0.025$.

    a. 2.365

    b. 2.447

    c. 3.707

    d. 3.499

41. Find the critical value for an F statistic with 20 and 25 degrees of freedom and $\alpha = 0.05$.

    a. 1.72

    b. 2.01

    c. 2.07

    d. 2.70

42. Which measurement scale has meaningful intervals and data that can be ordered but no meaningful zero point?

    a. Ratio

    b. Interval

    c. Ordinal

    d. Nominal

43. Employees are issued identification numbers based on a combination of numbers and letters. The IDs are an example of which measurement scale?

    a. Ratio

    b. Interval

    c. Ordinal

    d. Nominal

44. In which measurement scale do values have an order, but differences between adjacent values are indeterminant?

    a. Ratio

    b. Interval

c. Ordinal

d. Nominal

45. Which measurement scale has intervals between adjacent scale values that are equal with respect to the attribute being measured and a true zero?

a. Ratio

b. Interval

c. Ordinal

d. Nominal

46. Temperature measured in Fahrenheit is an example of which measure scale?

a. Ratio

b. Interval

c. Ordinal

d. Nominal

47. When setting up a data collection scheme for a project, which data set will be the most useful for measuring sensitivity to process changes?

a. Discrete

b. Continuous

c. Locational

d. Categorical

48. Compared to continuous data, discrete data requires what sample size to achieve the same Type II error?

a. Smaller

b. Equal

c. Larger

d. Cannot determine from the information given

49. In developing a data collection plan, what elements are needed?

a. What will be measured, training plan, who will measure, and the sampling plan

b. What will be measured, how it will be measured, who will measure, and the sampling plan

c. What will be measured, measurement cost, who will measure, and the sampling plan

d. What will be measured, process flow, who will measure, and the control plan

50. In a data collection plan, when defining "how to measure," which of the following elements should be considered?

a. How many to measure

b. Type of measure

c. Stratification plan

d. Who will take the measurement

51. In developing a data collection plan, what is included in the sampling plan?

a. Method

b. Who

c. How many

d. Measure type

52. Which data collection method can be used to avoid repetitive recording of numbers and errors due to fatigue in manual data capture?

a. Data coding

b. Data auditing

c. Check sheets

d. Stratified sampling

53. A quality inspector tracks the occurrences of different types of defects. Which tool would be best for data collection?

a. Stem-and-leaf plot

b. Check sheet

c. Scatter plot

d. Histogram

54. When working with data that have been coded, what does decoding entail?

a. Classifying the data into an easily understood system

b. Turning the code back to actual measurements

c. Establishing confidentiality of the information

d. Using a deciphering device to read the hidden meaning

55. Which of the following sampling methods should be used when there is a mixture of parts from different machines, different streams, different raw material lots, or different process settings, and there is no homogeneity in the lot?

a. Random sampling

b. Sequential sampling

c. Stratified sampling

d. Biased sampling

56. The formula shown below is the calculation for:

$$\sqrt{\frac{\Sigma(x-\bar{x})^2}{n-1}}$$

a. sample standard deviation.

b. population standard deviation.

c. sample mean.

d. population mean.

57. The data value that appears most frequently in the sample is called the:

a. mean.

b. median.

c. mode.

d. range.

58. Which of the following terms is a measure of dispersion?

a. Mean

b. Range

c. Median

d. Mode

59. A Green Belt wants to determine if two quantitative variables are related. She would likely collect data and create a:

a. scatter plot.

b. histogram.

c. Pareto chart.

d. stem-and-leaf plot.

60. The figure below is an example of a:

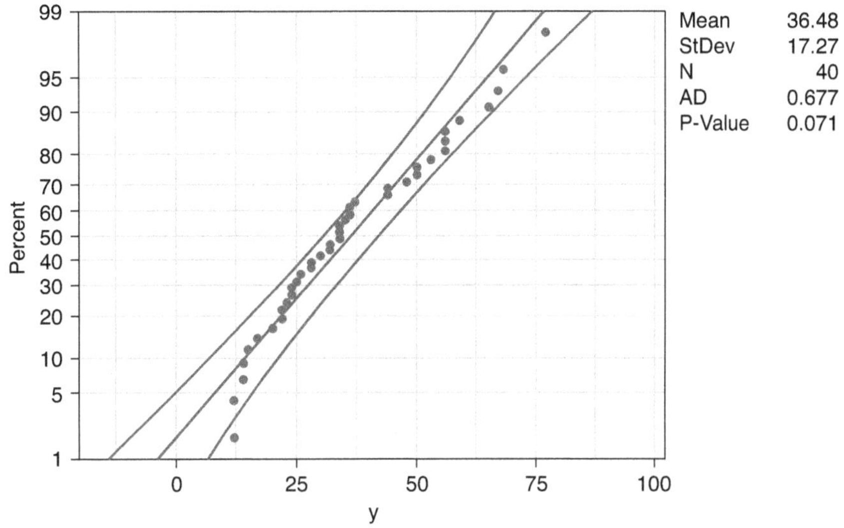

a. stem-and-leaf plot.

b. scatter plot

c. probability plot

d. box plot.

61. An advantage of a stem-and-leaf plot over a histogram is that the stem-and-leaf plot allows the analyst to:

a. re-create the original data.

b. determine the correlation between variables.

c. identify the largest driver of a problem.

d. study the shape of the data distribution.

62. A histogram is a graphical representation of a:

a. process flow.

b. frequency table.

c. value stream.

d. regression model.

Use the following figure to answer questions 63 and 64:

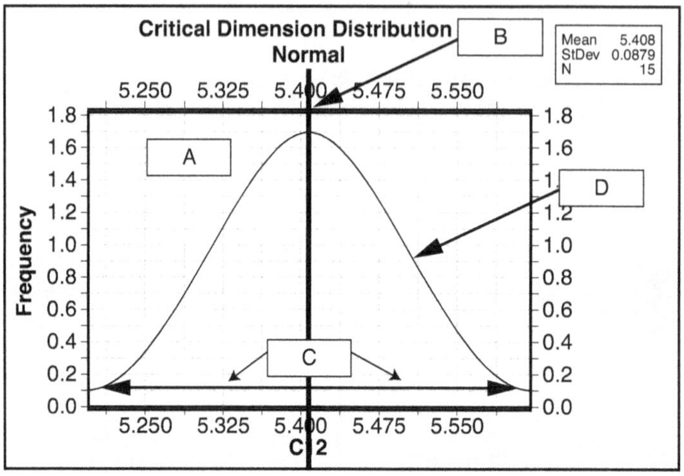

63. The graphical representation of spread is best identified by:
    a. A.
    b. B.
    c. C.
    d. D.
64. The graphical representation of shape is best identified by:
    a. A.
    b. B.
    c. C.
    d. D.
65. Which graphical tool is used to estimate the correlation coefficient, or make predictions using a regression line fitted to the data?
    a. Stem-and-leaf plot
    b. Scatter diagram
    c. Frequency distribution
    d. Run chart

66. Which graphical tool is useful to quickly identify any repeated data within a class interval?

 a. Stem-and-leaf plot
 b. Scatter diagram
 c. Frequency distribution
 d. Run chart

Use the box plot below to answer questions 67 and 68.

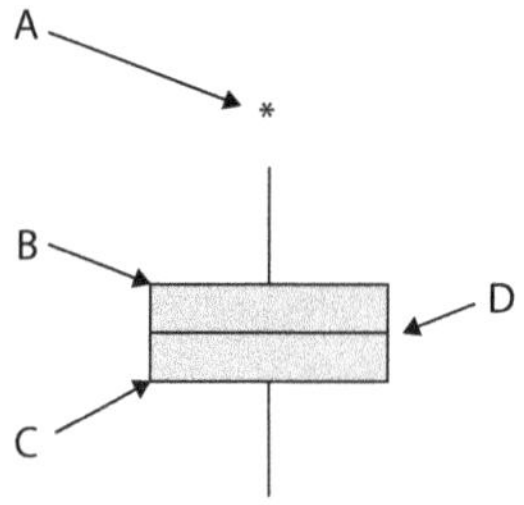

67. Which letter denotes the third quartile (Q3) of the sample?

 a. A.
 b. B.
 c. C.
 d. D.

68. Which letter denotes the median?

 a. A
 b. B
 c. C
 d. D

69. What is the MSA term for drift in absolute value over time?

 a. Calibration
 b. Stability
 c. Repeatability
 d. Reproducibility

70. What is the MSA term for closeness of repeated measurements to each other?

   a. Accuracy

   b. Stability

   c. Precision

   d. Reproducibility

71. The MSA term for accuracy of measurement at various measurement points of the measuring range in the equipment is:

   a. linearity.

   b. bias.

   c. normality.

   d. precision.

72. The MSA term for closeness to the true value or an accepted reference value is:

   a. linearity.

   b. bias.

   c. accuracy.

   d. precision.

73. What is the MSA term for drift in average measurements of an absolute value?

   a. Calibration

   b. Bias

   c. Accuracy

   d. Precision

74. Conducting an MSA should be the first step in:

   a. setting up an SPC process.

   b. conducting a DOE.

   c. developing control plans.

   d. preparing an FMEA.

75. When selecting samples to use in an MSA study, which factor is the most important to a successful outcome?

a. Randomly picking parts from a production bin

b. Picking consecutively produced parts from the process

c. Finding rejected parts that will be scrapped

d. Handpicking samples that cover the process spread

Use the following figure to answer questions 76–78.

Gage Repeatability and Reproducibility Data Collection Sheet											
Appraiser/ trial #	Part										Average
	1	2	3	4	5	6	7	8	9	10	
A1	23	33	22	10	15						
A2	21	31	21	10	16						
A3	19	29	23	10	11						
Average											$\overline{X}_a$
Range											$\overline{R}_a$

76. What is the rounded value for the overall average?

a. 18

b. 19

c. 20

d. 21

77. What is the value for R-bar?

a. 21.0

b. 21.5

c. 22.0

d. 23.0

78. What is the average reading for appraiser A2?

a. 18.4

b. 19.6

c. 19.8

d. 20.6

79. If the bias of a measurement system increases across the measurement range, the measuring device is said to have poor:

a. stability.

b. linearity.

c. bias.

d. reproducibility.

80. A measurement system has been evaluated with the following results: repeatability = 0.81 and reproducibility = 5.15. What is the gage R&R?

a. 2.44

b. 5.21

c. 5.96

d. 27.14

81. A precision-to-tolerance ratio is calculated to be 0.50. What can be said about the measurement system?

a. The measurement system is stable.

b. The variability of the measurement system is too large.

c. The variability of the products is too large.

d. The specification range is too small.

82. What is the measurement system concept depicted in the figure below?

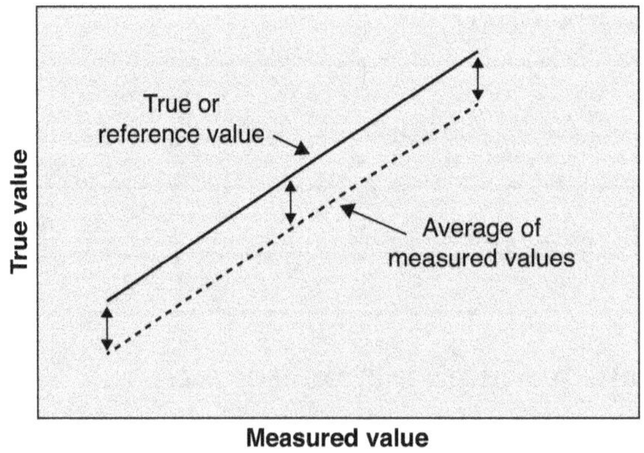

a. Precision

b. Stability

c. Linear bias

d. Repeatability

83. What components comprise the total variation of a measurement system?

a. Gage R&R and appraiser variation

b. Part-to-part variation and gage R&R

c. Equipment variation and gage R&R

d. Part-to-part variation and repeatability

84. What method can be used to assess a measurement system that uses pass/fail classification?

a. AIAG methodology

b. ANOVA analysis

c. Control chart method

d. Percent agreement

85. An attribute agreement analysis can include output on:

a. each appraiser versus standard and part variation.

b. all appraisers versus the standard and between-appraiser agreement.

c. within-appraiser agreement and part variation.

d. all appraisers versus standard, and gage R&R.

86. What is the total variation of the measurement system that has part variation equal to 1.60 and gage R&R equal to 0.89?

a. 1.83

b. 2.49

c. 2.65

d. 3.35

87. The purpose of calibration is to:

a. ensure that the measurement system is performing correctly.

b. vet the measurement standard against the equipment.

c. calculate the gage R&R of an instrument.

d. adjust measurement devices to reduce defect detection.

88. When conducting an MSA study, a common error made while performing GR&R is:

    a. using parts that cover the range of the gage.

    b. not randomizing the samples during measurement.

    c. using overtrained inspectors.

    d. using parts that cover the specification range.

89. What is the term for measurements taken simultaneously with multiple measurement devices of the same type for parts coming from multiple manufacturing streams?

    a. Measurand

    b. Calibration

    c. Measurement system error

    d. Measurement correlation

90. When the process variation is significantly smaller than the width of the specification, the process is said to be:

    a. out of spec.

    b. capable.

    c. Six Sigma capable.

    d. centered on target.

91. The traditional definition of natural process limits is:

    a. $\pm 3\ \sigma$.

    b. $\pm 4\ \sigma$.

    c. $\pm 5\ \sigma$.

    d. $\pm 6\ \sigma$.

92. The term for the ability of the process to meet the expected specifications is:

    a. process performance.

    b. process capability.

    c. process control.

    d. process spread.

93. The capability index Cr is the inverse of:

a. Cp.

b. Cpk.

c. Pp.

d. Ppk.

94. The capability index calculated using the within-subgroup standard deviation is:

a. Cp.

b. X-bar.

c. Pp.

d. Ppk.

95. Which scenario is best suited for using capability process measures (CPMs)?

a. A process with only short-term fluctuations

b. A process with no historical data available

c. A process with inconsistent variations

d. A process producing a consistent output within specifications

96. The performance location of a nonstable process is denoted by:

a. Cp.

b. Cpk.

c. Pp.

d. Ppk.

Use the following table for questions 97–100.

Sample size	125
Subgroup size	5
# subgroups	25
X-double bar	0.52
R-bar	0.66
Sample std deviation	0.29

USL	2
LSL	0
UCLx	0.90
LCLx	0.13
UCLr	1.40
LCLr	0

97. Estimate the value of Cp.

a. 0.60

b. 0.61

c. 1.11

d. 1.17

98. Estimate the value of Cpk.

a. 0.60

b. 0.61

c. 1.11

d. 1.17

99. Calculate the estimated value of Pp.

a. 0.60

b. 0.61

c. 1.15

d. 1.17

100. Calculate the estimated value of Ppk.

a. 0.60

b. 0.61

c. 1.11

d. 1.17

101. If a process is centered at specification target, what will be true?

a. Cp = Cpk

b. Cp > Cpk

c. Cp = Pp

d. Cpk = Ppk

102. A capability metric that uses a target value is known as the:

a. Cp.

b. Cpk.

c. Cpm.

d. Ppk.

103. A normally distributed process metric is currently running at a 3-sigma level of quality. What is the long-term yield of the process?

a. 93.32%

b. 95.45%

c. 99.73%

d. 99.99966%

SOLUTIONS

1. a; A process map shows a set of interrelated activities that transform inputs into outputs. [III.A]
2. d; The first step in creating a process map is to identify the boundaries of the process of interest. The other three steps will be taken as you progress through the project. [III.A]
3. b; Work instructions (WIs) are more granular descriptions of a process. In addition to the information contained in procedures, WIs also include who does what and how it is done. [III.A]
4. d; A *method* is an unwritten process that must be followed consistently. Both the FMEA and WI are written documented processes. The product map is not a tool used in SS. [III.A]
5. c; Measles charts show the physical location of defects. [III.A]
6. d; The Pareto chart helps narrow the team's focus by differentiating between the vital few and trivial many causes of a problem. This principle is also called the 80–20 rule. [III.A]
7. a; The six sides have equal probability. Thus $P(X=2)=1/6$. [III.B.1]
8. b; The probability of Event A or B is found using the addition rule: $P(A)+P(B)-P(A \& B)$. [III.B.1]
9. b; Using the addition rule, $P(4 \text{ or } 6)=P(4)+P(6)-P(4 \& 6)=1/6+1/6-0=1/3$. [III.B.1]
10. a; Using the general addition rule, $P(A)+P(B)-P(A \& B)$. If Events A and B are mutually exclusive, $P(A \text{ and } B)=0$. Therefore, the formula reduces to $P(A \text{ or } B) = P(A)+P(B)$. [III.B.1]
11. b; Here, we are given P(Machine A and Defective) = 0.02 and P(Machine A) = 0.40. We are asked to find the conditional probability P(Defective|Machine A). Let P(A) = P(Machine A) and P(B) = P(Defective). Then, using the conditional probability formula:

$$P(B \mid A)=\frac{P(A \text{ and } B)}{P(A)}=\frac{P(\text{Machine A and Defective})}{P(\text{Machine A})}=\frac{0.02}{0.40}=0.05.$$

[III.B.1]

12. b; There are a total of 100 complaints. Fifteen concerned wrong connectors. The probability that a randomly chosen complaint is a wrong connector complaint is 15/100. [III.B.1]

13. c; There are two types of shipping complaints: shipping damage and lost shipment. The total number of these complaints is $23+10=33$. Therefore, P(Shipping complaint) = 33/100. [III.B.1]

14. c; The probability that a connector complaint concerns missing connectors is found using the condition probability formula. Let P(Connector complaint) = P(A) and P(Missing connector) = P(B). We then can calculate:

$$P(B \mid A) = \frac{P(A \text{ and } B)}{P(A)} = \frac{32/100}{47/100} = \frac{32}{47}.$$

[III.B.1]

15. c; Five reports will be chosen from 15, in no particular order. We can use the combination formula to solve using $n=15$ and $r=5$.

$$nCr = \frac{n!}{r!(n-r)!} = \frac{15!}{5!(15-5)!} = \frac{15 \times 14 \times 13 \times 12 \times 11}{5 \times 4 \times 3 \times 2 \times 1} = 3003$$

[III.B.1]

16. d; Three candidates will be chosen for different positions, so order matters here. Using the permutation formula with $n=20$ and $r=3$ yields:

$$nPr = \frac{n!}{(n-r)!} = \frac{20!}{(20-3)!} = 20 \times 19 \times 18 = 6840.$$

[III.B.1]

17. d; The central limit states that averages tend to be normally distributed, even if the original distribution is not normally distributed. [III.B.2]

18. c; This is an application of the central limit distribution. If samples are taken from a distribution with mean μ and standard deviation σ, then the distribution of sample averages will approximately follow a normal distribution with mean μ and standard deviation $\sigma/\sqrt{n}$, where n is the number of data points used to calculate a sample average. Here, we are given that $\mu=10$, $\sigma=10$, and $n=36$. Therefore, the sample average will follow a normal distribution with mean 10 and standard deviation $10/\sqrt{36}=1.667$. [III.B.2]

19. a; Since control charts like X-bar and R charts and X-bar and s charts are plotted with averages of the individual readings, the charts are fairly robust against departures from normality. This is a consequence of the central limit theorem. [III.B.2]

20. c; The population is the collection of all items under consideration. [III.B]

21. d; Greek letters denote population parameters. Latin letters are generally used to denote sample statistics and, sometimes, parameters. [III.B]

22. a; The normal probability density function is $f(x)=\frac{1}{\sigma\sqrt{2\pi}}e^{-\frac{(x-\mu)^2}{2\sigma^2}}$. [III.C]

23. d; The Poisson probability mass function is $P(X=x)=\frac{\lambda^x e^{-\lambda}}{x!}$. [III.C]

24. b; The Poisson distribution's mean and variance are equal to λ. [III.C]

25. a; The normal distribution is fully described by two parameters: μ and σ^2. [III.C]

26. c; The Poisson distribution is used to count defects per unit. The number of defects has no natural upper limit. [III.C]

27. b; The binomial distribution is used to find the probability of x defectives in a sample of n. [III.C]

28. a; The Student's t distribution is symmetric and has a heavier tail than the normal distribution. As the sample size n increases, the tails of the Student's t distribution become less heavy, and the distribution looks more and more normal. [III.C]

29. d; We test the ratio of two variances using the F distribution. [III.C]

30. b; The chi-square distribution is used to test a variance or standard deviation against a standard. [III.C]

31. c; This is an application of the binomial distribution with $n=6$, $p=0.10$. To find P(X = 1) we use:

$$\mathrm{P}(X=1)=\binom{n}{x}p^x(1-p)^{n-x}$$

$$=\binom{6}{1}0.10^1(1-0.10)^{6-1}=6\times 0.10\times 0.90^5=0.3543.$$

[III.C]

Section III Solutions

32. c; This is an application of the Poisson distribution in which we are counting defects per unit. If there are an average of 0.01 defects per unit, we should expect (0.01)(100) = 1 defect per 100 units. We can use the Poisson distribution with $\lambda = 1$. To find the probability of finding one or zero defects in 100 units:

$$P(X = 0 \text{ or } X = 1) = \sum_{x=0}^{1} \frac{e^{-\lambda}\lambda^{x}}{x!} = \frac{e^{-1}1^{0}}{0!} + \frac{e^{-1}1^{1}}{1!} = 2e^{-1} = 0.7358.$$

[III.C]

33. a; The Z transformation standardizes any normal distribution with a mean μ and variance σ^2 into a standard normal with mean equal to 0 and variance equal to 1. Note:

$$Z = \frac{X - \mu}{\sigma}.$$

[III.C]

34. b; Among the choices, only the Student's t distribution has negative values. [III.C]

35. b; To find $P(1.00 < Z < 2.50)$, we can use a standard normal table to find $P(Z < 2.50) - P(Z < 1.00) = 0.99379 - 0.84134 = 0.15245$. See the sketch below.

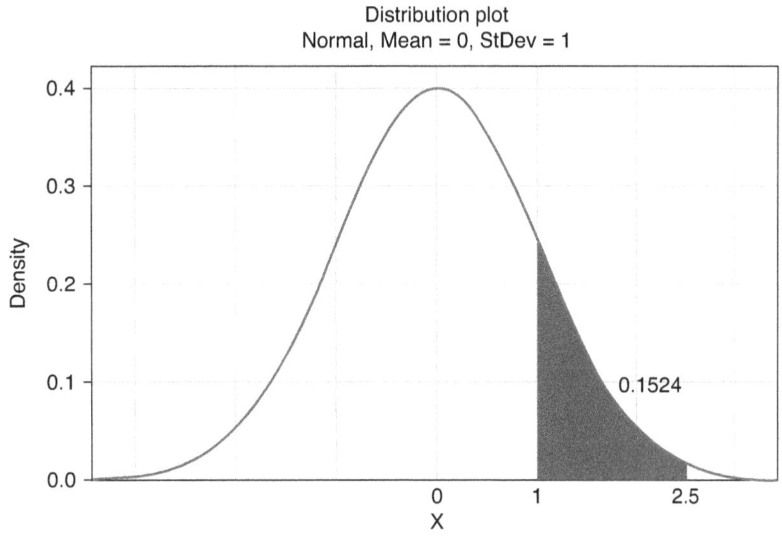

[III.C]

36. a; The weight variable follows a $N(16.0, 0.005^2)$ distribution. To find $P(X > 16.0075)$, we will use the standard normal distribution.

$$P(X > 16.0075) = P\left(Z > \frac{X - \mu}{\sigma}\right) = P\left(Z > \frac{16.0075 - 16.0}{0.005}\right) = P(Z > 1.5)$$

Using a cumulative standard normal table, we find $P(Z > 1.5) = 1 - P(Z < 1.5) = (1 - 0.9332) = 0.0668 = 6.68\%$. Note that the standard normal table in Appendix F of *The ASQ Certified Six Sigma Green Belt Handbook* gives values for $P(Z > z)$, so the answer can be taken directly from the table. [III.C]

37. b; To find $P(-1.33 < Z < 2.75)$, we can use a standard normal table to find $P(Z < 2.75) - P(Z < -1.33) = 0.99702 - 0.09176 = 0.90526$. [Depending on the standard normal table you are using, you may have to find $P(Z < -1.33)$ by calculating $1 - P(Z < 1.33) = 1 - P(Z < 1.33) = 1 - 0.90824 = 0.09176$.] See the sketch below.

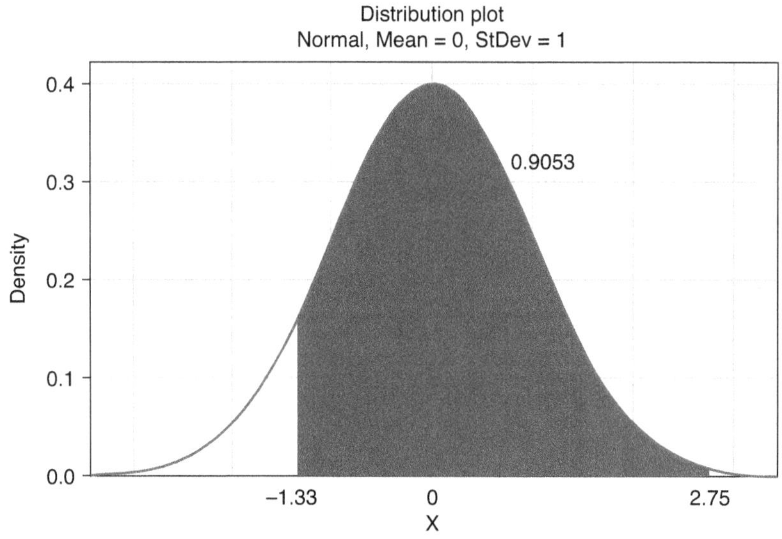

[III.C]

38. a; To find $P(Z \geq 0.50)$, we can find $1 - P(Z < 0.50)$ using the standard normal table, in which $1 - P(Z < 0.50) = 1 - 0.69146 = 0.30854$. Note that the standard normal table in Appendix F of *The ASQ Certified Six Sigma Green Belt Handbook* gives values for $P(Z > z)$, so the answer can be taken directly from the table. [III.C]

39. c; Using the chi-square table, we look up the value associated with 5 degrees of freedom and $\alpha = 0.05$ to get 11.070. Appendix I in *The ASQ Certified Six Sigma Green Belt Handbook* can be used as a reference. [III.C]

40. a; Using the Student's *t* table, we look up the value associated with 7 degrees of freedom and $\alpha = 0.025$ to get 2.365. Appendix L in *The ASQ Certified Six Sigma Green Belt Handbook* can be used as a reference. [III.C]

41. b; Using the F table, we look up the value associated with 20 and 25 degrees of freedom and $\alpha = 0.05$ to get 2.01. Appendix G in *The ASQ Certified Six Sigma Green Belt Handbook* can be used as a reference. [III.C]

42. b; Data on the interval scale have intrinsic numerical meaning, can be ordered, and the intervals between values are meaningful. However, interval data does not have a true zero point. Examples include temperature in Fahrenheit and Celsius, and calendar years. [III.D.1]

43. d; Data on the nominal scale have no intrinsic numerical meaning and no ordering scheme. Often, nominal data is made up of words or numerical sequences that serve as labels, such as social security numbers or employee IDs. [III.D.1]

44. c; In the ordinal measurement scale, data have inherent numerical value and can be ordered, but the difference between values is not necessarily equal. Ranks are an example of the ordinal measurement scale. [III.D.1]

45. a; The ratio measurement scale has inherent numerical value, can be ordered, has known intervals between data points, and has a true zero. Examples are weights, length, costs, or pressure readings. [III.D.1]

46. b; Temperature measured in degrees Fahrenheit is an example of the interval scale. The values have intrinsic numerical meaning, can be ordered, and the intervals between values are meaningful. However, interval data does not have a true zero point. Both the Fahrenheit and the Celsius scales have arbitrary zero points. [III.D.1]

47. b; Continuous data contains the most information and is more sensitive to process changes than discrete data. [III.D.1]

48. c; Discrete data do not contain as much information as continuous data. To achieve the same Type II error, a larger sample size must be collected compared to continuous data. [III.D.1]

49. b; A data collection plan will include the elements: what will be measured, how it will be measured, who will measure, and the sampling plan. [III.D.2]

50. c; The stratification plan and any needed data tags should be a consideration when defining how to measure. [III.D.2]

51. c; How many samples should be taken and where and how often are all elements of the sampling plan. [III.D.2]

52. a; Data coding can be used to avoid repetitive recording of numbers and errors due to fatigue in manual data capture. [III.D.2]

53. b; A check sheet would be the natural choice for tracking the frequency of defect types. [III.D.2]

54. b; Decoding data in this context means returning it to actual measurements. [III.D.2]

55. c; Use stratified sampling when the population is made of different groups. We stratify by group and then take a random sample within each group. [III.D.2]

56. a; The formula for the sample standard deviation, or s, is equal to $\sqrt{\frac{\Sigma(x-\bar{x})^2}{n-1}}$. [III.D.3]

57. c; The mode is the most frequently occurring value in the data set. There may be one mode, more than one mode, or no mode at all. [III.D.3]

58. b; The range is a measure of dispersion. [III.D.3]

59. a; Use a scatter plot to study the relationship between two quantitative variables. [III.D.4]

60. c; The plot shown is an example of a probability plot. [III.D.4]

61. a; Both the stem-and-leaf plot and the histogram show the shape of a data distribution. However, unlike the histogram, the stem-and-leaf plot allows the analyst to re-create the original data. [III.D.4]

62. b; A histogram is a column chart that depicts a frequency distribution. [III.D.4]

63. c; The spread is depicted by the width of the distribution, shown by C in the graph. [III.D.4]

64. d; In the graph, the shape of the distribution is shown by D. [III.D.4]

65. b; A scatter diagram is used to study the relationship between two quantitative variables. It can be used to estimate the correlation coefficient or make predictions using a regression line fitted to the data. [III.D.4]

66. a; A stem-and-leaf plot is similar to a histogram, except that the data values can be discerned from the graph. [III.D.4]

67. b; The third quartile (Q3) of the data is shown by arrow B in the box plot. Seventy-five percent of the sample is below the Q3 line. [III.D.4]

68. d; The median of the sample, or Q2, is shown by the arrow D in the box plot. Half of the data is below this point. [III.D.4]

69. b; Stability is the term for drift in absolute value over time. Higher stability results in less shifting over time. [III.E]

70. c; Precision is the closeness of repeated readings to each other. [III.E]

71. a; Linearity is the accuracy of measurement at various measurement points of the measuring range in the equipment. [III.E]

72. c; Accuracy is closeness to the true value, or to an accepted reference value. [III.E]

73. a; Calibration is the MSA term for drift in average measurements of an absolute value. [III.E]

74. a; Conducting a measurement system analysis is a standard rule for using SPC. [III.E]

75. d; Handpicking samples that cover the process spread is critical to ensure that you are identifying the correct process variation. In fact, ensuring sample parts that test the range of the gage is even better for looking at the gage variation. [III.E]

76. c; Using the table below, calculate the averages for each row (A1 = 20.6, A2 = 19.8, A3 = 18.4) and then calculate the X-bar for all three rows = 19.6. Use standard rounding rules: round to nearest whole number or round to even if a 0.5 occurs. [III.E]

Gage Repeatability and Reproducibility Data Collection Sheet											
Appraiser/ trial #	Part										
	1	2	3	4	5	6	7	8	9	10	Average
A1	23	33	22	10	15						
A2	21	31	21	10	16						
A3	19	29	23	10	11						
Average											$\overline{X}_a$
Range											$\overline{R}_a$

77. a; Calculate the range for each row (A1 = 23, A2 = 21, A3 = 19) and then calculate the R-bar = 21. [III.E]

78. c; Calculate only the line for A2 to find the average of the five parts measured = 19.8. [III.E]

79. b; If the bias of a measurement system increases across the measurement range, the measuring device is said to have poor linearity. A system with good linearity will have a constant bias across the range of the measurement. [III.E]

80. b; To calculate gage R&R, we use:

$$\sigma_{Gage\,R\&R} = \sqrt{\sigma^2_{repeatability} + \sigma^2_{reproducibility}}$$
$$= \sqrt{0.81^2 + 5.15^2}$$
$$= 5.21.$$

[III.E]

81. b; A precision-to-tolerance ratio (PTR) is a measure of the capability of the measurement system. PTR values less than 0.10 are considered good, while values greater than 0.30 require improvement in the measurement system in terms of reducing its variability. [III.E]

82. c; The figure depicts the concept of linear bias, in which the measured values are consistently biased across the range of measurement. [III.E]

83. b; Part-to-part variation and gage R&R comprise the total variation of the measurement system. [III.E]

84. d; Percent agreement can be used to assess a measurement system that uses pass/fail classification. [III.E]

85. b; The output for an attribute agreement analysis can include four components: each appraiser versus the standard, between appraisers, all appraisers versus the standard, and within appraiser. [III.E]

86. a; Total variation of the measurement system is a combination of gage R&R and part variation (PV). We calculate total variation using:

$$TV = \sqrt{GRR^2 + PV^2}$$
$$= \sqrt{0.89^2 + 1.60^2} = 1.83.$$

[III.E]

87. a; The purpose of calibration is to ensure that the measurement system is performing correctly. [III.E]

88. b; When conducting an MSA study, a common error made while performing GR&R is not randomizing the samples during measurement. [III.E]

89. d; Measurement correlation is used when measurements are taken simultaneously with multiple measurement devices of the same type for parts coming from multiple streams of manufacturing. [III.E]

90. b; When the process variation is significantly smaller than the width of the specification, the process is said to be capable. [III.F.1]

91. a; Traditionally, natural process limits are defined as $\pm 3\ \sigma$. For a normal distribution, this width maps to 99.73% of the distribution. [III.F.1]

92. b; The ability of the process to meet the expected specifications is process capability. [III.F]

93. a; Cr is the inverse of Cp. Cr is expressed as a percentage and shows the percentage of specification used up by the process variation. [III.F]

94. a; The Cp index is calculated using the within-subgroup standard deviation from a control chart. [III.F.3]

95. c; Cpm is appropriate for processes with inconsistent variations, as it helps assess the process's ability to consistently meet specifications despite variations. [III.F.2]

96. d; Ppk is an index for unstable processes that indicates a general location within the specification limits. [III.F.3]

97. d; Using the information from the control chart, we note that d_2 for $n=5$ is equal to 2.326. Refer to Appendix E of *The ASQ Certified Six Sigma Green Belt Handbook.* Then we estimate Cp:

$$\hat{C}_p = \frac{(USL - LSL)}{6\hat{\sigma}} = \frac{(USL - LSL)}{6\left(\frac{\bar{R}}{d_2}\right)} = \frac{(2-0)}{6\left(\frac{0.66}{2.326}\right)} = 1.17.$$

[III.F.3]

98. b; Using the information from the control chart, we note that d_2 for $n=5$ is equal to 2.326. Refer to Appendix E of *The ASQ Certified Six Sigma Green Belt Handbook.* Then we estimate Cpk:

$$\hat{C}_{pk} = \min\left[\frac{(USL - \bar{\bar{X}})}{3\hat{\sigma}}, \frac{(\bar{\bar{X}} - LSL)}{3\hat{\sigma}}\right]$$

$$= \min\left[\frac{(USL - \bar{\bar{X}})}{3\left(\frac{\bar{R}}{d_2}\right)}, \frac{(\bar{\bar{X}} - LSL)}{3\left(\frac{\bar{R}}{d_2}\right)}\right]$$

$$= \min\left[\frac{(2-0.52)}{3\left(\frac{0.66}{2.326}\right)}, \frac{(0.52-0)}{3\left(\frac{0.66}{2.326}\right)}\right]$$

$$= \min\left[\frac{(2-0.52)}{3\left(\frac{0.66}{2.326}\right)}, \frac{(0.52-0)}{3\left(\frac{0.66}{2.326}\right)}\right]$$

$$= \min[1.74, 0.611] = 0.611.$$

[III.F.3]

Section III Solutions

99. c; To estimate Pp, we use:

$$\hat{P}_p = \frac{(USL - LSL)}{6\hat{\sigma}} = \frac{(USL - LSL)}{6s} = \frac{(2-0)}{6 \times 0.29} = 1.15.$$

[III.F.3]

100. a; To estimate the value of Ppk, we use:

$$\hat{P}_{pk} = \min\left[\frac{(USL - \bar{\bar{X}})}{3s}, \frac{(\bar{\bar{X}} - LSL)}{3s}\right]$$
$$= \min\left[\frac{(2-0.52)}{3 \times 0.29}, \frac{(0.52-0)}{3 \times 0.29}\right]$$
$$= \min[1.70, 0.60]$$
$$= 0.60.$$

[III.F.3]

101. a; If the process is centered at the specification target, the Cpk value will equal Cp. [II.F.3]

102. c; The Cpm capability measure is used when a process has a target value. [IIIF.3]

103. a; A 3-sigma level process indicates that the specifications of a normally distributed process distribution map to ± 3 standard deviations for the mean. The short-term yield is 99.73%. If we apply a 1.5 standard deviation shift in the mean, the long-term yield reduces to 93.32%. This can be determined using a chart that maps sigma level, to short and long-term DPMO and yield. [III.F.4]

Section IV
Analyze Phase

(Exam Targets 18 Questions)

A. Exploratory data analysis

1. Multi-vari studies

Select appropriate sampling plans to create multi-vari study charts and interpret the results for positional, cyclical, and temporal variation. (Create)

2. Correlation and linear regression

Describe the difference between correlation and causation. Calculate the correlation coefficient and linear regression and interpret the results in terms of statistical significance (*p*-value). Use regression models for estimation and prediction. (Evaluate)

B. Hypothesis testing

1. Basics

Distinguish between statistical and practical significance. Determine appropriate sample sizes and develop tests for significance level, power, and type I and type II errors. (Apply)

2. Tests for means, variances, and proportions

Conduct hypothesis tests to compare means, variances, and proportions (paired-comparison t-test, F-test, analysis of variance (ANOVA), chi-square) and interpret the results. (Analyze)

C. Additional analysis methods

1. Gap analysis

Analyze scenarios to identify performance gaps and compare current and future states using pre-defined metrics. (Analyze)

2. Root cause analysis

Use cause-and-effect diagrams, relational matrices, Five Whys, fault tree analysis, and other problem-solving tools to identify the true cause of a problem. (Analyze)

QUESTIONS

1. The multi-vari chart is a useful tool for analyzing what types of variation?
   a. Average, spread, and location
   b. Skewness and kurtosis
   c. Cyclical, temporal, and positional
   d. Thermal and geographic placement
2. In a multi-vari chart study, another term for shift-to-shift variation is:
   a. positional variation.
   b. cyclical variation.
   c. normal variation.
   d. temporal variation.
3. When developing a multi-vari sampling plan, what should be the first step?
   a. Create a value stream process flow map of the target area.
   b. Select the process and the characteristics to be investigated.
   c. Record the time and values from each sample set in a table format.
   d. Use a cross-functional team to conduct a root cause analysis of the target area.

Metal flanges are produced on two machines. Readings of a critical dimension were taken from flanges from the two machines across three shifts. Refer to the following multi-vari chart to answer questions 4–6.

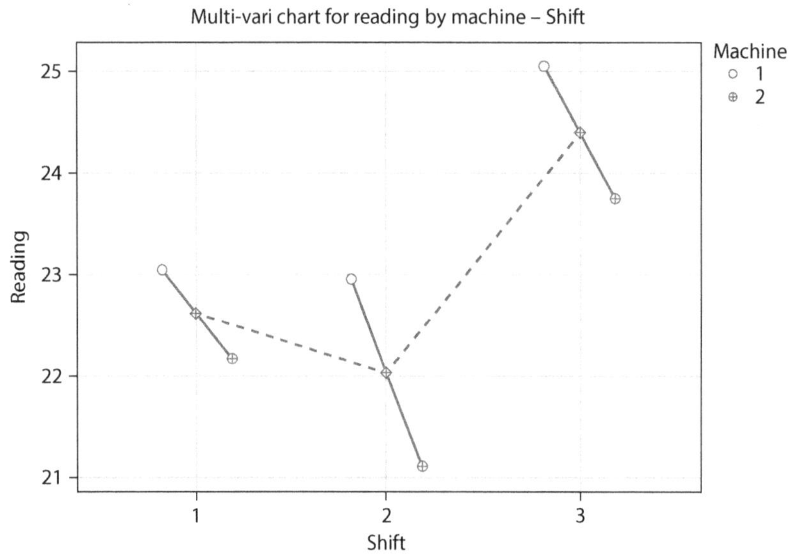

4. Which shift has the smallest variation between readings from the two machines?
   a. Shift 1
   b. Shift 2
   c. Shift 3
   d. All shifts have the same variability.
5. Which source of variation is the largest?
   a. Cyclic
   b. Temporal
   c. Positional
   d. Reading-to-reading
6. Which shift has the highest mean reading?
   a. Shift 1
   b. Shift 2
   c. Shift 3
   d. Cannot be determined using the information given

Four operators measured the inner diameter of a plastic sheath in three locations. Refer to the following multi-var chart to answer questions 7–9.

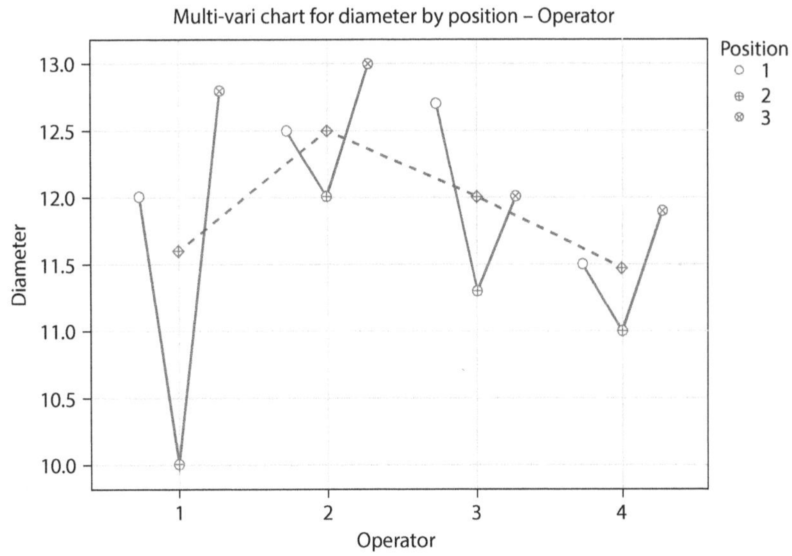

7. Which position on the sheath has the smallest mean diameter?
   a. Position 1
   b. Position 2
   c. Position 3
   d. Cannot be determined from the information given
8. Which operator has the largest variation in diameter readings?
   a. Operator 1
   b. Operator 2
   c. Operator 3
   d. Operator 4
9. The dotted line represents average diameter readings across:
   a. positions.
   b. departments.
   c. shifts.
   d. parts.
10. An event that gives rise to another event is an example of:
    a. correlation.
    b. independent variable.
    c. causation.
    d. dependent variable.
11. If, during a study, a relationship between two or more sets of data is found, then we say that there is a(n):
    a. dependent variable.
    b. independent variable.
    c. causation.
    d. correlation.

Use the following figure for questions 12 and 13.

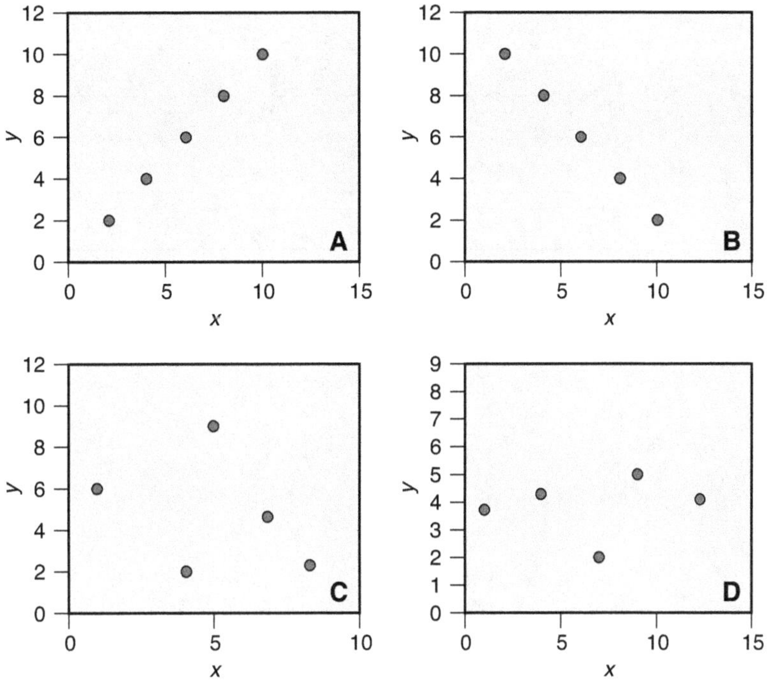

12. Which graph shows a strong positive correlation?
    a. A
    b. B
    c. C
    d. D
13. Which graph shows a strong negative correlation?
    a. A
    b. B
    c. C
    d. D

Use the following figure to answer questions 14–16. As shown, the (x, y) pairs are (2, 9), (7, 14), and (9, 13).

x values	y values	$(x-\bar{x})(y-\bar{y})$
2	9	
7	14	
9	13	
$\bar{x}=$	$\bar{y}=$	$\Sigma(x-\bar{x})(y-\bar{y})=$

14. The values for the x-bar and y-bar, respectively, are:

    a. 6, 12.

    b. 7, 13.

    c. 8, 13.

    d. 9, 14.

15. The value for $\Sigma(x-\bar{x})(y-\bar{y})$ is equal to:

    a. 4.

    b. 5.

    c. 14.

    d. 17.

16. Calculate the sample linear correlation using the data in the table.

    a. 0.09

    b. 0.59

    c. 0.89

    d. 1.78

17. An estimated simple linear regression line is represented by:

    a. $x=b_0+b_1y+e$.

    b. $\hat{y}=b_1+b_0x$.

    c. $\hat{y}=b_0+b_1x$.

    d. $y=\beta_0+\beta_1x_1+\beta_2x_2+\varepsilon$.

18. What is the interpretation of slope in a simple linear regression?

    a. If $\hat{y}$ increases by 1 unit, x will increase by the amount of the slope.

    b. If x increases by 1 unit, $\hat{y}$ will change by the amount of the slope.

    c. If x equals 0, $\hat{y}$ will equal the slope.

    d. An increase in x will cause $\hat{y}$ to increase.

19. A simple linear regression model was estimated to predict the shipping cost for an item based on the number of units: $\hat{y} = -3.45 + 12.5x$. What is the expected shipping cost for 100 units?

    a. $357.50

    b. $1246.55

    c. $1250.00

    d. $1253.45

20. For the fitted regression model $\hat{y} = -12.45 + 2.75x$, the value –12.45 represents:

    a. the y-intercept.

    b. the slope.

    c. the dependent variable.

    d. the independent variable.

21. The diagram below shows the critical region for a two-tailed hypothesis test. The total value of the shaded areas is equal to:

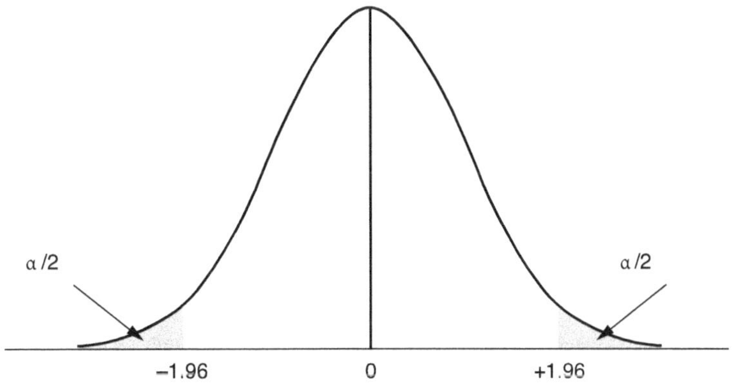

    a. 0.025.

    b. 2.50.

c. 0.05.

d. 0.005.

22. Which action can decrease the potential of a Type II error?

a. Increasing sample size

b. Decreasing alpha level

c. Increasing variability

d. Decreasing the difference to detect

23. In hypothesis testing, when the null hypothesis is false, as the effect size increases:

a. Type I error increases.

b. Type II error decreases.

c. Type I error decreases.

d. Type II error decreases.

24. The complement of Type II error is:

a. power.

b. significance level.

c. confidence level.

d. effect size.

25. When the null hypothesis is true, changing the significance level from 0.05 to 0.01 will ________ the probability of a Type I error.

a. increase

b. decrease

c. not affect

d. eliminate

26. Calculate the sample size needed to assure a margin of error of 10 for a 95% two-sided confidence interval on the mean. The population variance is equal to 1200.

a. 24

b. 47

c. 236

d. 55,320

27. A Type II or beta error occurs when the null hypothesis:

    a. is rejected when it is true.

    b. is not rejected when it is true.

    c. is rejected when it is false.

    d. is not rejected when it is false.

28. Which of the following sets of hypotheses is appropriate to test the difference between two means?

    a. $H_0 : \mu_d = 0$ *vs.* $H_1 : \mu_d \neq 0$

    b. $H_0 : \mu_1 - \mu_2 = 0$ *vs.* $H_1 : \mu_1 - \mu_2 \neq 0$

    c. $H_0 : \mu_1 - \mu_2 \neq 0$ *vs.* $H_1 : \mu_1 - \mu_2 = 0$

    d. $H_0 : \bar{x}_1 - \bar{x}_2 = 0$ *vs.* $H_1 : \bar{x}_1 - \bar{x}_2 \neq 0$

29. If the ratio of two sample variances from normal distributions exceeds the critical value of F at a defined level of significance:

    a. the null hypothesis is rejected.

    b. the null hypothesis is accepted.

    c. the null hypothesis is not rejected.

    d. the null hypothesis is significant.

30. With n observations from a random sample, each of which could be classified into exactly one of k categories, how many degrees of freedom does a goodness-of-fit test with specified probabilities have?

    a. N

    b. $n-1$

    c. $k-1$

    d. $k-2$

31. A training company has shown that its study program results in a significantly higher mean score on a 100-point certification exam than those who did not use the program. Further investigation shows that the mean score for those who used the training program was 80.4, compared to 80.2 for those who did not. This is an example of a:

    a. false claim.

    b. result being statistically insignificant.

c. result being statistically insignificant but practically significant.

d. result being statistically significant but not practically significant.

32. A two-sided hypothesis test on two means is being conducted. Each sample comes from a symmetric distribution, and the sample size of each group is equal to 5. The variances of the groups are assumed to be equal. What is the critical value for the test if $\alpha = 0.01$?

a. $t_{0.005,9} = 3.250$

b. $t_{0.005,8} = 3.355$

c. $t_{0.01,9} = 2.821$

d. $t_{0.01,8} = 2.896$.

33. A new electric vehicle is tested for its maximum range under a single charge. Four vehicles were tested on a track with the following mileage results: {260, 252, 254, 254}. Assuming mileage is normally distributed, test whether the average range is at least 250 miles using an alpha = 0.05.

a. $z_0 = 2.89$; reject H_0.

b. $z_0 = 2.89$; do not reject H_0.

c. $t_0 = 2.89$; reject H_0.

d. $t_0 = 2.89$; do not reject H_0.

34. A process is running at a 5.0% defective rate. In a pilot test of a potential improvement, 225 items were processed with 5 defectives. Has the improvement significantly improved the defective rate? Test at $\alpha = 0.05$.

a. $Z_0 = -1.93$, the defective rate has not significantly improved.

b. $Z_0 = -1.93$, the defective rate has significantly improved.

c. $t_0 = -1.93$, the defective rate has not significantly improved.

d. $t_0 = -1.93$, the defective rate has significantly improved.

35. When calculating the 95 percent confidence interval for the mean of a sample of 256 customers, the average order is \$80, and the population standard deviation is \$16, so the range of the interval is:

a. $79.88 \leq \mu \leq 80.12$.

b. $79.02 \leq \mu \leq 80.98$.

c. $78.04 \leq \mu \leq 81.96$.

d. $76.08 \leq \mu \leq 83.92$.

36. To compare the means of three or more groups, which of the following tests is the most appropriate?

    a. Student's *t*-test

    b. Fisher's exact test

    c. Chi-squared test

    d. ANOVA

37. A confidence interval on the standard deviation of a normal distribution uses critical values from which distribution?

    a. Chi-square

    b. Poisson

    c. Binomial

    d. Standard normal

38. When calculating a 90 percent confidence interval for a proportion with a sample size of 100 and a sample proportion of 0.33, the interval is calculated using:

    a. 0.33 +/− 1.645 (0.047).

    b. 0.33 +/− 1.645 (0.0022).

    c. 0.33 +/− 1.96 (0.047).

    d. 0.33 +/− 1.96 (0.0022).

39. A Six Sigma Green Belt is studying the output of two milling machines. Assuming the output of each machine follows a normal distribution, test whether there is a significant difference in the variances of the two machines at $\alpha = 0.10$. Use the information in the following table.

Machine 1	Machine 2
$n = 20$	$n = 15$
$s^2 = 4.55$	$s^2 = 1.55$

    a. $F_0 = 2.94$, $F_{crit} = 1.92$, Reject H_0.

    b. $F_0 = 2.94$, $F_{crit} = 1.97$, Reject H_0.

    c. $F_0 = 2.94$, $F_{crit} = 2.33$, Reject H_0.

    d. $F_0 = 2.94$, $F_{crit} = 2.40$, Reject H_0.

Participants in a Green Belt course were given tests before and after a training session. The instructor wants to determine if the training was effective. Use the results in the following table to answer questions 40–42.

Participant	Pre-test	Post-test
1	55	80
2	67	70
3	80	90
4	79	83
5	84	88

40. What is the appropriate set of hypotheses for the test?

   a. $H_0: \mu_1 = \mu_2$ *vs.* $H_1: \mu_1 \neq \mu_2$

   b. $H_0: \mu_1 - \mu_2 = 0$ *vs.* $H_1: \mu_1 - \mu_2 < 0$

   c. $H_0: \mu_D = 0$ *vs.* $H_1: \mu_D \neq 0$

   d. $H_0: \mu_D = 0$ *vs.* $H_1: \mu_D < 0$

41. What is the value of the test statistic?

   a. $t_0 = -0.99$

   b. $t_0 = -1.44$

   c. $t_0 = -1.99$

   d. $t_0 = -2.22$

42. What is the critical value for the test, assuming $\alpha = 0.05$?

   a. $t = 1.812$

   b. $Z = 1.860$

   c. $Z = 2.015$

   d. $t = 2.132$

Use the following ANOVA table to answer questions 43–46.

Source	SS	df	MS	F
Supplier	360	4		
Error	400	20		
Total	760	23		

43. Which set of hypotheses is correct for the analysis?

a. H_0: $\mu_1 = \mu_2 = \mu_3 = \mu_4$ vs. H_1: $\mu_1 \neq \mu_2 \neq \mu_3 \neq \mu_4$

b. H_0: $\mu_1 = \mu_2 = \mu_3 = \mu_4$ vs. H_1: At least one mean is not equal to the rest.

c. H_0: $\mu_1 = \mu_2 = \mu_3 = \mu_4 = \mu_5$ vs. H_1: $\mu_1 \neq \mu_2 \neq \mu_3 \neq \mu_4 \neq \mu_5$

d. H_0: $\mu_1 = \mu_2 = \mu_3 = \mu_4 = \mu_5$ vs. H_1: At least one mean is not equal to the rest.

44. What is the calculated value of the F statistic?

a. 0.22

b. 0.90

c. 4.50

d. 9.00

45. What is the critical value from the F table if $\alpha = 0.01$?

a. 2.25

b. 3.8

c. 4.43

d. 14.02

46. What is the conclusion for the statistical test?

a. $F_{calc} > F_{crit}$, Reject H_0.

b. $F_{calc} < F_{crit}$, Reject H_0.

c. $F_{calc} > F_{crit}$, Do not reject H_0.

d. $F_{calc} < F_{crit}$, Do not reject H_0.

A Green Belt conducted a chi-square goodness of fit test to determine if the current defect rates differed from historical percentages. Use the table below to answer questions 47–49.

	Probability	Observed frequency	Expected frequency
Paint run	0.16	27	33.28
Paint blister	0.28	60	58.24
Decal crooked	0.42	100	87.36
Door cracked	0.14	21	29.12
Total			208.00

47. What is the calculated value of the chi-square statistic?
    a. –0.29
    b. 0.46
    c. 5.33
    d. 6.25
48. What is the critical value of the chi-square goodness of fit test at $\alpha = 0.05$?
    a. 5.991
    b. 7.815
    c. 9.348
    d. 11.143
49. What conclusion can be stated, assuming $\alpha = 0.05$?
    a. Reject H_0, defect percentages have changed.
    b. Do not reject H_0, defect percentages have not changed.
    c. Reject H_0, defect percentages have not changed.
    d. Accept H_0, defect percentages have not changed.
50. Contingency tables use the chi-square procedure to test whether there is a(n):
    a. association between two classifications.
    b. difference in population means.
    c. difference in population variances.
    d. difference in the observed frequency compared to historical proportions.
51. Which distribution is used for a hypothesis test on a proportion?
    a. Poisson
    b. Chi-square
    c. Normal
    d. Student's *t*
52. Which of the following statements represents the correct scenarios to identify performance gaps?
    a. Reactive: Customer satisfaction before having launched a product/service
    b. Proactive: Measuring operational efficiency after the organizational change

c. Proactive: Operational efficiency prior to rolling out an organizational change

d. Reactive: Customer perception prior to a product/service launch

53. A proactive approach to identifying performance gaps when an organization wants to identify opportunities for improvement would be to examine:

    a. customer perception prior to a product or service launch.

    b. customer satisfaction after having launched a product/service.

    c. operational efficiency after the organizational change.

    d. results of the last internal management system audit.

54. A gap analysis in Six Sigma is most useful during which phase?

    a. Measure

    b. Analyze

    c. Improve

    d. Control

55. Which statement correctly distinguishes fault tree analysis from Five Whys analysis?

    a. Fault tree analysis is a top-down failure analysis in which a failure of a system is analyzed using a combined series of lower-level events, while Five Whys diagnoses the failure by asking Why repeatedly until the root cause is determined.

    b. Fault tree analysis is a bottom-up failure analysis in which a failure of lower-level events is analyzed using system failures, while Five Whys diagnoses the failure by repeatedly asking Why until the root cause is determined.

    c. Five Whys analysis is a top-down failure analysis in which a failure of a system is analyzed using a combined series of lower-level events, while fault tree analysis diagnoses the failure by asking Why repeatedly until the root cause is determined.

    d. The two terms are consistent and interchangeable, as both are applied toward root cause analysis.

56. The analytical method that is mainly used in reliability engineering to understand how systems can fail is:

    a. FMEA.

    b. fault tree analysis.

c. cause-and-effect diagram.

d. root cause analysis.

57. Which methodology takes into consideration multiple causes for each effect and assigns a logical relationship between the causes?

a. FMEA

b. Fault tree analysis

c. Cause-and-effect diagram

d. Root cause analysis

58. Which of the following tools can be used to drill down to a root cause?

a. Eight Wastes

b. 5S

c. SMED

d. 5 Whys

59. Which of the following tools is used to provide visual identification of many potential causes of a problem?

a. Decision tree

b. Process flow chart

c. PERT chart

d. Check sheet

60. A persistent questioning technique used to probe deeper and reach the cause of a problem is known as the:

a. scatter diagram.

b. cause-and-effect diagram.

c. relationship diagram.

d. Five Whys.

61. The tool that is a top-down failure analysis in which a system failure is analyzed using a combined series of lower-level events is called:

a. FMEA.

b. fault tree analysis.

c. cause-and-effect diagram.

d. RACI.

62. A factor that caused a nonconformance and should be permanently eliminated through process improvement is the:

a. root cause.

b. cause-and-effect diagram.

c. relationship diagram.

d. Five Whys.

63. A tool for analyzing process dispersion is the:

a. root cause.

b. cause-and-effect diagram.

c. relationship diagram.

d. Five Whys.

64. A management tool that depicts the interactions among factors in a complex situation is the:

a. root cause.

b. cause-and-effect diagram.

c. interrelationship diagram.

d. Five Whys.

65. A technique for discovering the root causes of a problem and showing the relationship of those causes is the:

a. root cause.

b. cause-and-effect diagram.

c. relationship diagram.

d. Five Whys.

66. What values are summed up to form the top-level cause within fault tree analysis?

a. The AND/OR logic leading to the failure of the system

b. The logical relationship between the causes

c. The nominal quantity of identified causes

d. The probabilities of each assignable cause

67. The labels on the main bones of a cause-and-effect diagram refer to:

a. problems or goals.

b. categories of possible causes.

c. defect rates.

d. individual workers.

Use the following diagram to answer questions 68 and 69.

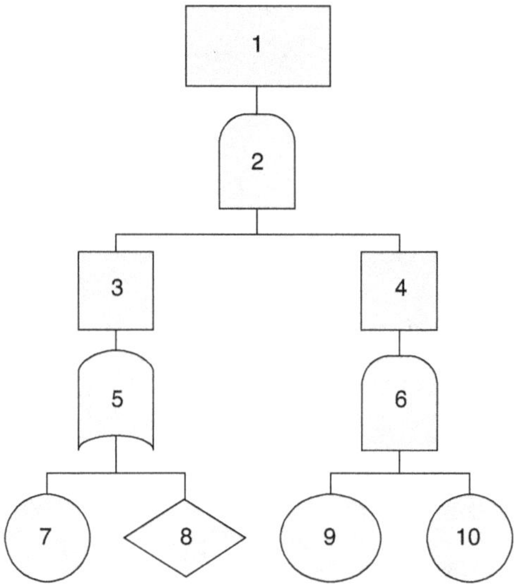

68. Which of the FTA symbols indicates an AND gate?

a. 2 and 6

b. 3 and 4

c. 5

d. 7, 9, and 10

69. Which of the FTA symbols indicates the primary event?

a. 1

b. 3

c. 4

d. 3 and 4

Use the diagram below to answer questions 70 and 71.

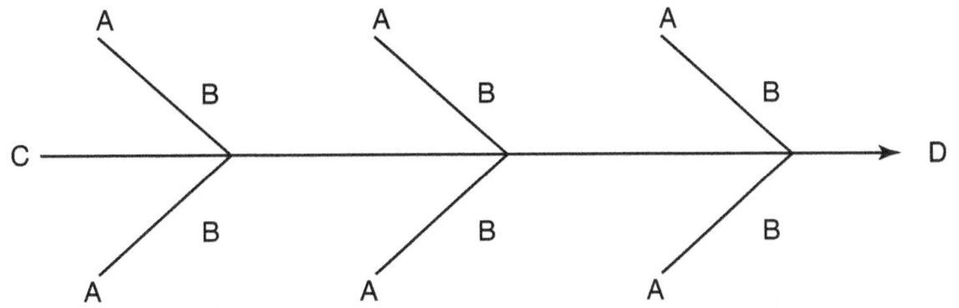

70. Categories such as Machinery and Materials are placed in which position on the cause-and-effect diagram?

    a. A

    b. B

    c. C

    d. D

71. The problem that is being addressed is placed in what position of the diagram?

    a. A

    b. B

    c. C

    d. D

SOLUTIONS

1. c; The multi-vari chart is a useful tool for analyzing cyclical, temporal, and positional variation. [IV.A.1]
2. d; Temporal variation is another term for shift-to-shift variation. [IV.A.1]
3. b; The first step when developing a multi-vari sampling plan is to select the process and the characteristics to be investigated. [IV.A.1]
4. a; The first shift has the smallest difference between machine readings. [IV.A.1]
5. b; The shift-to-shift, or temporal, variation is the largest. [IV.A.1]
6. c; Shift 3 has the highest mean reading, as shown in the dotted line. [IV.A.1]
7. b; Position 2 is the smallest reading for all operators. [IV.A.1]
8. a; Operator 1 has the largest differences in readings across the shaft. [IV.A.1]
9. a; The dotted line shows the average diameter readings across the three positions of the shaft. [IV.A.1]
10. c; Causation is a reason that produces an effect that gives rise to any action or condition. [IV.A.2]
11. d. Correlation is a relationship between two or more sets of data found during a study. [IV.A.2]
12. a; Graph A with the upward slope shows a strong positive correlation. [IV.A.2]
13. b; Graph B with the downward slope shows a strong negative correlation. [IV.A.2]
14. a; The values for $\bar{x}$ and $\bar{y}$, respectively, would be 6 and 12. [IV.A.2]
15. d; The calculations are as follows:

	x	y	$(x-\bar{x})$	$(y-\bar{y})$	$(x-\bar{x})(y-\bar{y})$
	2	9	4	3	12
	7	14	1	2	2
	9	13	3	1	3
Sum	18	36			17
Average	6	12			

[IV.A.2]

16. c; To calculate the sample correlation coefficient, we can use:

$$r = \frac{\Sigma(x-\bar{x})(y-\bar{y})}{(n-1)s_x s_y}.$$

Using a calculator, we can find the sample standard deviation of each variable: $s_x = 3.61$ and $s_y = 2.65$. The numerator of the formula is equal to 17, as shown in the preceding table. Note that we have three (x, y) pairs, so $n = 3$. Then,

$$r = \frac{\Sigma(x - \bar{x})(y - \bar{y})}{(n-1)s_x s_y} = \frac{17}{2 \times 3.61 \times 2.65} = 0.89. \text{ [IV.A.2]}$$

17. c; The estimated simple linear regression line is represented by the formula $\hat{y} = b_0 + b_1 x$. [IV.A.2]

18. b; The interpretation of the slope is that if x increases by 1 unit, $\hat{y}$ will change by the amount of the slope. [IV.A.2]

19. b; Using the simple linear regression equation, we calculate:

$$\hat{y} = -\ 3.45 + 12.5 \times 100 = 1246.55.$$

[IV.A.2]

20. a; The value –12.45 is the y-intercept of the regression line. [IV.A.2]

21. c; The value of each shaded area is 0.025 for a two-tailed test, for a total of 0.05. [IV.B.1]

22. a; To reduce the potential of a Type II error, we can increase the sample size in our study. [IV.B.1]

23. b; If the null hypothesis is false, the alternative hypothesis is true. If the effect size is large, we are more likely to (correctly) reject the null. This means that Type II error decreases as the effect size increases. Note that if the alternative hypothesis is true, the only error we can make is a Type II. [IV.B.1]

24. a; Power is equal to (1–β), so it is the complement of Type II error. [IV.B.1]

25. b; If the null hypothesis is true, the only type of error we can make is a Type I or alpha error. If the significance level is changed from 0.05 to 0.01, more evidence will be required to reject the null. Therefore, the probability of a Type I error will decrease. [IV.B.1]

26. b; We are given $E = 10$ and $\sigma^2 = 1200$. Let $Z = 1.96$ (based on a 95% two-sided confidence interval). Then, using the formula for sample size:

$$n = \frac{Z^2\sigma^2}{E^2} = \frac{1.96^2 \times 1200}{10^2} = 46.10 \uparrow 47.$$

Note that we always round the result up to the next highest integer value. [IV.B.1]

Section IV Solutions

27. d; A Type II error occurs when we fail to reject the null when the alternative is true. See the following truth table.

	Decision	
Truth	Do not Reject H_o	Reject H_o
H_o	Correct	Type I, alpha error
H_1	Type II, beta error	Correct

[IV.B.1]

28. b; The null hypothesis always contains some form of equality, and hypotheses are always written in terms of population parameters. To test two means that we can use:

$$H_0: \mu_1 - \mu_2 = 0 \text{ vs. } H_1: \mu_1 - \mu_2 \neq 0. \text{ [IV.B.2]}$$

29. a; The null hypothesis rejection area is equal to or greater than the critical value of F at a defined significance level, so if the test statistic is greater than the critical value, the null hypothesis is rejected. [IV.B.2]

30. c; For a chi-squared goodness-of-fit test, there will be k–1 degrees of freedom, where k is the number of categories. [IV.B.2]

31. d; This is an example of a result that is statistically significant but not practically significant. [IV.B.1]

32. b; The degrees of freedom are equal to 5 + 5–2 = 8. Since the test is two-sided, we use $\alpha/2 = 0.01/2 = 0.005$. From the Student's t table, the critical value for t(0.005, 8) = 3.355. [IV.B.1]

33. c; We set up the one-sided hypotheses for the mean:

$$H_0 : \mu = 250$$
$$H_1 : \mu > 250$$

Because we do not know the population standard deviation, we will use the Student's t distribution for this test.

$$\bar{X} = \frac{\sum x_i}{n} = \frac{260 + 252 + 254 + 254}{4} = 255$$

$$s = \sqrt{\frac{\sum x_i^2 - n\bar{x}^2}{n-1}} = \sqrt{\frac{260^2 + 252^2 + 254^2 + 254^2 - 4 \times 255^2}{4-1}} = 3.46$$

$$t_0 = \frac{254 - 250}{4.73/\sqrt{4}} = 2.89$$

We will reject H_0 if $t_0 > t(0.05, 3) = 2.353$. Since $2.89 > 2.353$, we reject H_0 and conclude that the average mileage is greater than 250. [IV.B.2]

34. b; This is a one-sided test since we are trying to prove that the improvement decreases the defect rate. The hypotheses we are testing are:

$$H_0 : p = 0.05$$

$$H_1 : p < 0.05$$

We note that $n = 225$, $p_0 = 0.05$, and $\hat{p} = 0.022$ and use the normal approximation method:

$$z_0 = \frac{(\hat{p} - p_0)}{\sqrt{\frac{p_0(1 - p_0)}{n}}} = \frac{(0.022 - 0.05)}{\sqrt{\frac{0.05(1 - 0.05)}{225}}} = -1.93.$$

We reject H_0 if $z_0 < -z_{0.05}$. From the table, we find $-z_{0.05} = -1.645$. Since -1.93 is less than -1.645, we reject H_0. The improvement has reduced the rate of defects. [IV.B.2]

35. c; The sample mean is 80, the population standard deviation is equal to 16, and $n = 256$. Using the confidence interval formula for the mean with a known variance:

$$\bar{X} \pm Z\frac{\sigma}{\sqrt{n}}$$

$$80 \pm 1.96\frac{16}{\sqrt{256}}$$

$$80 \pm 1.96\frac{16}{16}$$

$$[78.04, 81.96].$$

[IV.B.2]

36. d; To compare the means of three or more groups, we use analysis of variance (ANOVA). [IV.B.2]

37. a; The chi-square distribution is used to test a standard deviation or variance from a normal distribution. [IV.B.2]

38. a; The Z value for a two-tailed 90 percent confidence interval is 1.645. The estimated proportion is 0.33, and $n = 100$. We then use the formula:

$$\hat{p} \pm Z\sqrt{\frac{\hat{p}(1-\hat{p})}{n}}$$

$$0.33 \pm 1.645\sqrt{\frac{0.33(0.67)}{100}}$$

$$0.33 \pm 1.645\sqrt{\frac{0.33(0.67)}{100}}$$

$$0.33 \pm 1.645 \times 0.047.$$

[IV.B.2]

39. b; The hypotheses for this test are:

$$H_0 : \frac{\sigma_1^2}{\sigma_2^2} = 1$$

$$H_1 : \frac{\sigma_1^2}{\sigma_2^2} > 1$$

The test statistic is calculated by:

$$F_0 = \frac{s_1^2}{s_2^2} = \frac{4.55}{1.55} = 2.94.$$

Noting that $n_1 = 20$ and $n_2 = 15$, we calculate the degrees of freedom for the F critical value of 19 and 14. Looking up the table value for $F(0.10, 19{,}14)$, we get $F_{crit} = 1.97$. Since $F_0 > F_{crit}$, we reject H_0 and conclude that the variances of the machines are not equal. [IV.B.2]

40. d; This is a paired sample test in which we want to show that the training succeeded. This will be the case if the mean difference between the individuals' pre- and post-test scores is negative. Thus, the appropriate hypotheses for this study are:

$$H_0\text{: } \mu_D = 0 \text{ vs. } H_1\text{: } \mu_D < 0.$$

[IV.B.2]

41. d; We first calculate the mean difference and the standard deviation of the differences using the table:

Participant	Pre-test	Post-test	Difference
1	55	80	–25
2	67	70	–3
3	80	90	–10
4	79	83	–4
5	84	88	–4
Mean difference			–9.2
Std deviation			9.257

Then we calculate:

$$t_0 = \frac{\bar{x} - \mu_0}{s_d / \sqrt{n}} = \frac{-9.2 - 0}{9.26 / \sqrt{5}} = -2.22.$$

[IV.B.2]

42. d; The degrees of freedom for the Student's t distribution for this test are equal to 4, since there are 5 students. Noting that this is a one-sided study, we find $t(0.05, 4) = 2.132$. Since $|-2.22| > 2.132$, we reject H_0 is $|\ t_0\ | > t_{crit}$ and conclude that the training was effective. [IV.B.2]

43. d; Since the degrees of freedom for Supplier are equal to 4, we know we have 5 means that we are comparing. The alternative hypothesis for ANOVA is that at least one mean is not equal to the rest. [IV.B.2]

44. c; The completed ANOVA table is shown:

Source	SS	df	MS	F
Supplier	360	4	90	4.5
Error	400	20	20	
Total	760	23		

MS is equal to SS/df for Supplier and Error. To calculate the F statistics, we divide $MS_{supplier}$ by MS_{Error}. [IV.B.2]

45. c; The critical value $F(4, 20, 0.01) = 4.43$ can be found using the statistical tables. [IV.B.2]

46. a; The calculated $F = 4.50$ is greater than the table value $F(4, 20, 0.01) = 4.43$. Therefore, we reject H_0 and conclude that at least one supplier mean is not equal to the rest. [IV.B.2]

47. c; To calculate the chi-square statistic, we use:

$$\chi^2 = \Sigma \frac{(O_i - E_i)^2}{E_i}$$

$$= \frac{(27-33.28)^2}{33.28} + \frac{(60-58.24)^2}{58.24} + \frac{(100-87.36)^2}{87.36} + \frac{(21-29.12)^2}{29.12}$$

$$= 1.185 + 0.0532 + 1.829 + 2.264$$

$$= 5.331.$$

[IV.B.2]

48. b; The critical value is 7.815. The degrees of freedom are equal to one less than the number of categories, k. Here, $(k-1)=3$. Since goodness of fit is always a right-tailed test, the χ^2 table value at $\alpha=0.05$ value is applied. [IV.B.2]

49. b; Since the calculated chi-square value of 5.33 is not greater than the critical value of 7.815, we do not reject H_0. There is insufficient evidence to claim that the percentages have changed. [IV.B.2]

50. a; A contingency table shows two or more categorical variables with counts in each cell. The statistical test associated with a contingency table is the chi-square procedure. We test whether there is an association between the two categorical variables. For example, we could study whether defect types are associated with shift. [IV.B.2]

51. c; For the hypothesis test on proportions, we often use a normal approximation. [IV.B.2]

52. c; Proactive: Operational efficiency prior to rolling out an organizational change. [IV.C.1]

53. a; The question asks for a proactive approach, which indicates something prior to a launch. [IV.C.1]

54. c; Gap analysis is most useful during the improve phase of a Six Sigma project. [IV.C.1]

55. a; Fault tree analysis is a top-down failure analysis in which a failure of a system is analyzed using a combined series of lower-level events, while Five Whys diagnoses the failure by asking *why* repeatedly until the root cause is determined. [IV.C.2]

56. b; Reliability engineers use fault tree diagrams and analysis to uncover actual and potential system errors. [IV.C.2]

57. b; The fault tree process is able to look at various streams/roots that can lead to the issues being studied. [IV.C.2]

58. d; The Five Whys can be used to drill down to a root cause. [IV.C.2]

59. a; A decision tree provides visual identification of many potential causes of a problem. [IV.C.2]

60. d; The tool that is a persistent questioning technique used to probe deeper to reach the cause of a problem is known as the Five Whys, which is also called the 5W1H method. [IV.C.2]

61. b; Fault tree analysis is a top-down approach to analysis. [IV.C.2]

62. a; This question asks how to prevent reoccurrences, thus requiring a root cause. [IV.C.2]

63. b; The C&E diagram is used in the analysis of process dispersion. [IV.C.2]

64. c; The interrelationship diagram, also known as the relationship diagram, depicts interactions among the factors in a complex situation. [IV.C.2]

65. d; The tool described is the Five Whys, which is also called the 5W1H method. [IV.C.2]

66. d; For a fault tree analysis, the probability of each assignable cause is calculated in either series or parallel. All the probabilities are then summed up to the top-level cause. [IV.C.2]

67. b; In a cause-and-effect, or fishbone, diagram, the main bones are labeled with categories of possible causes of the problem. [IV.C.2]

68. a; Symbols 2 and 6 denote AND gates. The output for an AND gate occurs only if all inputs occur. [IV.C.2]

69. a; Symbol 1 represents the primary, or head, event. This is the failure mode being investigated. [IV.C.2]

70. a; The categories of causes such as manpower, materials, machinery, methods, and mother nature would be placed on the ends of the bones, in position A. [IV.C.2]

71. d; The problem being addressed (the effect) in a fishbone diagram is placed at the head of the fish, in position D. [IV.C.2]

Section V
Improve Phase

(Exam Targets 16 Questions)

A. DOE

1. Basic terms

Define and describe terms such as independent and dependent variables, factors and levels, responses, treatments, errors, repetition, blocks, randomization, effects, and replication. (Understand)

2. DOE graphs and plots

Interpret main effects analysis and interaction plots. (Apply)

B. Implementation planning

Apply implementation planning by using proof of concepts, try-storming simulations, and pilot tests. (Apply)

C. Lean Tools

1. Waste elimination

Select and apply tools and techniques for eliminating or preventing waste, including pull systems, kanban, 5S, standard work, and poka-yoke. (Apply)

2. Cycle-time reduction

Use various techniques to reduce cycle time (continuous flow, setup reduction, single-minute exchange of dies (SMED)). (Analyze)

3. Kaizen and kaizen blitz

Define and distinguish between these two methods and apply them in various situations. (Apply)

QUESTIONS

1. The settings of a factor in an experimental design are referred to as:
   a. levels.
   b. blocks.
   c. groupings.
   d. replication.
2. A fractional factorial experiment is used to:
   a. express factor levels in percentages adding up to 100 percent.
   b. investigate the primary factor while controlling the interactions and effects of other variables.
   c. increase the combinations of factors and levels required to be run.
   d. reduce the combinations of factors and levels required to be run.
3. Randomization of experimental runs is intended to reduce:
   a. special cause variation.
   b. common cause variation.
   c. main effects.
   d. interaction effects.
4. An experiment must be run over the course of two days. The days are not of experimental interest, but they are incorporated into the design. This is an example of:
   a. blocking.
   b. confounding.
   c. repetition.
   d. replication.
5. Experimental error is defined as:
   a. errors introduced by faulty experimentation.
   b. variability not explained by the experimental factors.
   c. errors introduced by using a fractional factorial experiment.
   d. a specific type of confounding.

6. A Green Belt runs an experiment and finds that the yield of a process is maximized when the machine temperature is set to 450° F and pressure is set to 40 psi. What should be the next step?

   a. Implement the changes and update the work instructions.

   b. Train the operators to use the new settings.

   c. Run a confirmation experiment.

   d. Perform a kaizen blitz on the line.

7. A Green Belt runs a 2^3 full-factorial experiment and produces two samples from each run. This is an example of:

   a. replication.

   b. repetition.

   c. blocking.

   d. confounding.

8. For main effects analysis, a higher number of replications results in:

   a. reduced randomization of the experiment.

   b. an increased probability of confounding.

   c. a better estimate of experimental error, reducing alpha risk.

   d. a better estimate of experimental error, reducing beta risk.

9. An experiment that runs each possible combination of factor levels so that all interactions can be estimated is known as a:

   a. Latin square.

   b. fractional factorial.

   c. full factorial.

   d. Taguchi design.

10. In an experiment, the independent variables represent the:

    a. responses for each run.

    b. factors varied in the experiment.

    c. experimental error.

    d. experimental effect.

11. The larger the experimental effect, the more likely that the effect is significant, meaning that the:
    a. null hypothesis will be rejected.
    b. null hypothesis will not be rejected.
    c. null hypothesis will be accepted.
    d. null hypothesis can be disregarded.
12. A 3^2 experiment has at least how many runs?
    a. 6
    b. 8
    c. 9
    d. 12
13. A half-fraction of a 2^5 experiment has how many runs?
    a. 5
    b. 8
    c. 16
    d. 32
14. Comparative, screening, and mixture are examples of:
    a. experimental objectives.
    b. residuals.
    c. simulation studies.
    d. balanced designs.

Use the following three-factor experiment and the corresponding responses to answer questions 15–17.

Run	A	B	C	Y
1	–	–	–	10
2	+	–	–	5
3	–	+	–	5
4	+	+	–	3
5	–	–	+	2
6	+	–	+	1
7	–	+	+	7
8	+	+	+	10

15. What type of experiment is this?
    a. 2^3
    b. 3^2
    c. 2×3
    d. 2^{4-1}
16. What is the main effect of factor A?
    a. –1.25
    b. 0
    c. 0.625
    d. 1.25
17. What is the main effect of Factor B?
    a. –1.25
    b. 0.875
    c. 1.25
    d. 1.75
18. The risks of confounding for a fractional factorial design are higher when:
    a. there are no interaction effects.
    b. there are no main effects.
    c. main effects are present.
    d. interaction effects are present.

19. The interaction plots for an experiment with three factors are shown below. Based on the graphs, which two-factor interaction is the strongest?

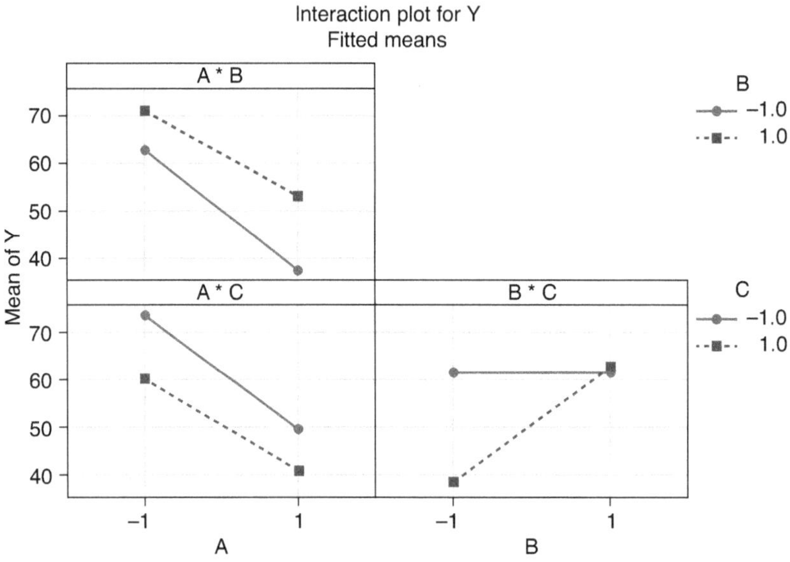

a. Interaction AB

b. Interaction AC

c. Interaction BC

d. Cannot be determined from the graph

20. A two-factor experiment was conducted. The analysis showed that the interaction effect was not significant. Based on the main effects plots below, how will the mean response be affected if Factor B is changed from the low level to the high level?

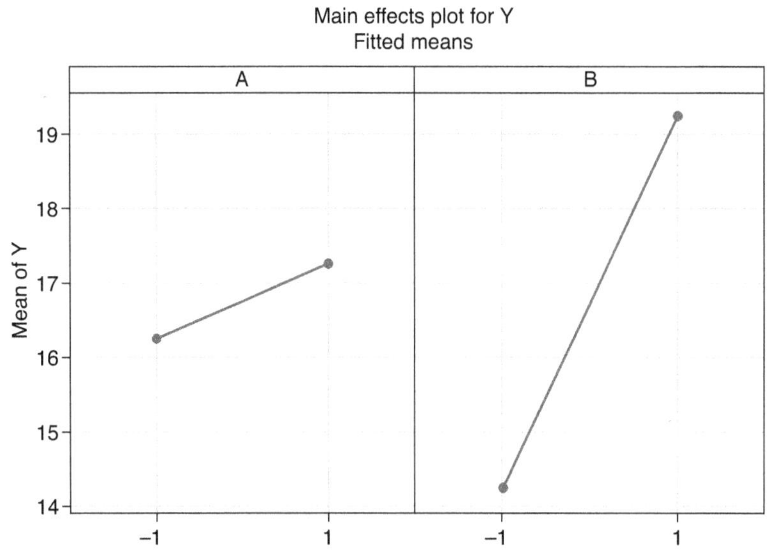

a. The mean response will decrease by 5 units.

b. The mean response will increase by 2.5 units.

c. The mean response will increase by 5 units.

d. The mean response will be dependent on the setting for Factor A.

21. Using the interaction graph below, which factor settings will result in the largest mean response?

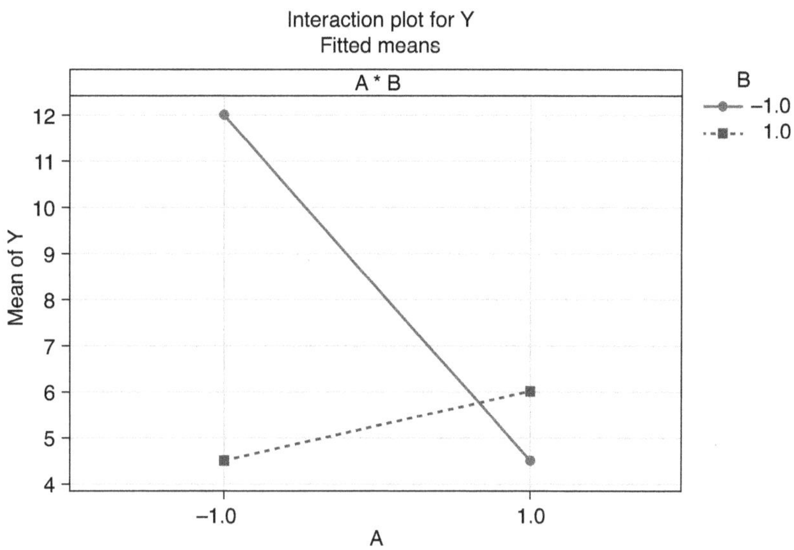

a. Factor A low, Factor B low

b. Factor A low, Factor B high

c. Factor A high, Factor B low

d. Factor A high, Factor B high

22. A technique used in early-stage design validation that confirms the feasibility of the initiative and presents it to the participants and stakeholders for the purpose of eliciting feedback and detailed inputs is called:

a. proof of concept.

b. try-storming simulations.

c. pilot test.

d. advanced product quality planning.

23. Try-storming is a technique that aims to:
    a. quickly simulate the most promising brainstorming ideas.
    b. move a Six Sigma team from the storming to norming stage.
    c. verify the effectiveness of the process solution.
    d. demonstrate that success criteria can be fulfilled.
24. A process that allows the team to sample the solution prior to completion is called:
    a. proof of concept.
    b. try-storming simulations.
    c. piloting testing.
    d. advanced product quality planning.
25. Select the option that correctly pairs the concept with its definition.
    a. Pilot tests: An early-stage design validation technique that confirms the feasibility of the initiative
    b. Proof of concept: Tests that are performed in the latter stages of the project to determine the readiness for broader implementation
    c. Try-storming simulations: Exploratory tests to incrementally build the concept toward completion and implementation
    d. Advance Product Quality Planning: Potential fulfillment of the needs analysis within business costs and constraints
26. Try-storming continues the work generated by brainstorming by which actions?
    a. Quickly simulating the most feasible ideas from the brainstorming session
    b. Critiquing the outputs of brainstorming and identifying potential defects and flaws for risk management
    c. Expanding the scope and reducing control of implementation activities
    d. Conducting extensive document reviews prior to simulation
27. What are the potential risks if the pilot test phase does not effectively restrict problematic solutions?
    a. The proof of concepts will be delayed, affecting the potential progress and readiness for completion.
    b. The try-storming will need to be canceled, pending a resolution of the pilot test.

c. The product malfunctions from previous versions could inhibit completion.

d. New defects and maladies could impair the benefits of the Six Sigma project.

28. The ideal relationship of lean and Six Sigma methods is best described as:

a. sequential; complete lean activities before starting Six Sigma.

b. mutually exclusive; deploy either lean or Six Sigma activities, but not both.

c. integrative; use both methods to reinforce and sustain improvements.

d. antagonistic; using lean methods will reduce the effect of Six Sigma.

29. The best way to reduce large, segregating inventories and carrying costs is to reduce:

a. dedicated full-time employees.

b. large work areas.

c. concurrent production of different models.

d. the quantity of work-in-process (WIP) inventory.

30. The 5S method is used to:

a. track process outcomes and identify out-of-control situations.

b. distinguish between special and common causes of variation.

c. create order and organization.

d. validate compliance to ISO 9001.

31. Which of the following tools is used to develop standard work?

a. Production capacity chart

b. Kanban

c. Control charts

d. 5S

32. On a production line, a kanban square is used to signal when another part is needed. This is an example of:

a. mistake-proofing.

b. a pull system.

c. standard work.

d. SMED.

33. An operator is tasked with placing workpieces into a milling machine. Once every so often, the operator places a piece into the machine upside down, resulting in a shutdown. What is the best way to address this problem?

    a. Retrain the operator

    b. Employ total productive maintenance

    c. Mistake-proof the machine

    d. Use a pull system for the process

34. Overproduction, excess motion, waiting, inventory level excess movement, defect correction, excess processing, and lost creativity are examples of:

    a. waste, or muda.

    b. capacity.

    c. value-added activities.

    d. Kano categories.

35. Any activity that does not add value to a product or service is classified as:

    a. waste.

    b. kanban.

    c. takt time.

    d. kaizen.

36. Defects, excess motion, and rework are all forms of:

    a. appraisal cost.

    b. the value stream.

    c. waste.

    d. pull systems.

37. Actions taken to eliminate waste (muda), unevenness (mura), and unreasonableness (muri) are known as:

    a. countermeasure activities.

    b. mistake and error-proofing (poka-yoke).

    c. the theory of constraint (TOC) approach.

    d. takt time analysis.

38. The crossover of Lean-Six Sigma objectives between traditional Six Sigma and traditional Lean is:

    a. shifted process average.

    b. reduced process variation.

    c. reduced cycle time.

    d. reduced process complexity.

39. The process that manages scheduling and material flow so that overproduction and excess inventories are eliminated is known as:

    a. a pull system.

    b. the visual factory.

    c. kanban.

    d. total productive maintenance.

40. Which advantages justify the importance of quick changeover or single-minute exchange of dies (SMED) implementation?

    a. Economic order quantity (EOQ) amortizes the costs of long setups

    b. Creates bottlenecks, drives up lot sizes, and reduces flexibility

    c. Mitigates the hidden costs of producing and transporting large amounts of inventory

    d. Reduces the time available to set up and produce a wider variety of products and services

41. What are examples of items that could be standardized to reduce internal setup times?

    a. Tools, fasteners, and placement

    b. Bottlenecks, lot sizes, and flexibility

    c. Inventory, shipment quantities, and receiving methods

    d. Six Sigma practices for define, measure, analyze, improve, and control phases

42. A strategy to reduce changeover time is to:

    a. reduce the number of activities performed while the machine is shut down.

    b. continue to run the machine to stockpile inventory.

    c. skip steps solely performed for safety.

    d. reduce training requirements for the set-up team.

43. The Gemba kaizen guidelines include which of the following major activities?

    a. Explaining project purpose, sending global correspondence, and instructing operators and fill ins

    b. Designing large-scale interventions and selecting kaizen plans

    c. Setting the target, selecting the leaders, checking the operating lines, and confirming the inventory

    d. Preparing standards and a summary report for every cross-function and division.

44. The main difference between kaizen and a kaizen blitz is the:

    a. focus on accuracy.

    b. pace at which improvements are made.

    c. emphasis on mistake-proofing.

    d. elimination of necessary movement.

SOLUTIONS

1. a; A level refers to the setting of a factor in an experimental design. [V.A.1]
2. d; Fractional factorials reduce the combinations of factors and levels for experiments. [V.A.1]
3. a; Randomization organizes experiments to lessen the effects of special cause variation by randomly assigning the order in which tests are run. [V.A.1]
4. a; Blocking variables account for additional variation added to the response from variables that are not of experimental interest. [V.A.1]
5. b; Experimental error is the variability not explained by the experimental factors. [V.A.1]
6. c; After analyzing an experiment, a confirmation test should be run using the new settings to check the consistency of the results. [V.A.1]
7. b; Repetition is used when multiple samples are created for the same run. Generally, the average and standard deviation of the samples are used in the analysis. [V.A.1]
8. d; A higher number of replications or a larger sample size improves the estimate of experimental error, reducing the beta risk. [V.A.2]
9. c; Full factorial experiments look at every possible combination in order to complete a full study of interactions. [V.A.2]
10. b; The independent variables are the factors that are varied in the experiment. [V.A.1]
11. a; When effects are significant, they are not purely the result of experimental error, and the null hypothesis is likely to be rejected. [V.A.1]
12. c; A 3^2 experiment has two factors varied at three levels each. The minimum number of runs for this experiment is 9. [V.A.1]
13. c; A half-fraction of a 2^5 experiment has $2^{5-1} = 2^4 = 16$ runs. [V.A.1]
14. a; Experimental objectives can be summarized under four categories: comparative, screening, response surface, and mixture. [V.A.1]
15. a; This experiment has three factors at two levels each, and every factor level combination is run. It is a 2^3, full-factorial experiment. [V.A.1]
16. a; The main effect of Factor A is calculated using:

$$m.e.(A) = \frac{A^+ - A^-}{4} = \frac{(5+3+1+10)-(10+5+2+7)}{4} = \frac{19-24}{4} = \frac{-5}{4} = -1.25.$$

[V.A.2]

17. d; The main effect for Factor B is calculated using:

$$m.e.(B)=\frac{B^{+}-B^{-}}{4}=\frac{(5+3+7+10)-(10+5+2+1)}{4}=\frac{25-18}{4}=\frac{7}{4}=1.75.\ [\text{V.A.2}]$$

18. d; If interaction effects are present, there is more of a risk of confounding in a fractional factorial experiment. [V.A.2]

19. c; The BC interaction is the strongest. The BC interaction plot shows lines that are the least parallel. Because there is a BC interaction, the effect of changing the B from the low to the high level cannot be determined unless the factor level setting of C is known, and vice versa. [V.A.2]

20. c; The mean response will increase by five units when Factor B is changed from the low to the high level. [V.A.2]

21. a; Setting both Factors A and B to their low levels will result in a mean response equal to 12. [V.A.2]

22. a; Proof of concept is an early-stage design validation technique that confirms the feasibility of the initiative. This is presented to the participants and stakeholders for eliciting feedback and detailed inputs. [V.B]

23. a; Try-storming is a technique that aims to quickly simulate the most promising brainstormed ideas. [V.B]

24. c; A pilot test is a process to hone the project prior to completion. There are opportunities to sample or improve the solution prior to full deployment. [V.B]

25. c; Try-storming simulations are exploratory tests to incrementally build the concept toward completion and implementation. [V.B]

26. a; In try-storming, the most feasible ideas from the brainstorming sessions are quickly simulated or developed for demonstration. [V.B]

27. d; The benefits of the Six Sigma project will be offset by the new risks, defects, and maladies introduced from the suboptimal implementation approach. [V.B]

28. c; The convergence of Lean approaches and Six Sigma methods will more efficiently deliver the intended benefits to cost, quality, delivery, customer satisfaction, and predictability. [V.C.1]

29. d; Lean reduces the WIP inventory, which reduces segregating inventories and carrying costs. [V.C.1]

30. c; 5S is used to create order and organization. [V.C.1]

31. a; A production capacity chart is used to develop standard work. [V.C.1]

32. b; Producing a piece only when it is needed by the next machine is an example of a pull system. [V.C.1]

33. c; The best way to address human errors that occur at random intervals is to mistake-proof the process. A fixture on the milling machine could be redesigned so that the piece will not fit if it is placed in the machine upside-down. [V.C.1]

34. a; Overproduction, excess motion, waiting, inventory level excess movement, defect correction, excess processing, and lost creativity are examples of waste. [V.C.1]

35. a; Any activity that does not add value to a product or service is classified as waste. [V.C.1]

36. c; Defects, excess motion, and rework are all forms of waste. [V.C.1]

37. a; Countermeasure activities eliminate waste (muda), unevenness (mura), and unreasonableness (muri). [V.C.1]

38. b; The crossover of Lean Six Sigma, between traditional Lean and traditional Six Sigma, is reduced process variation. [V.C.1]

39. a; A pull system is a process used to manage scheduling and material flow to avoid overproduction and excess inventories. [V.C.1]

40. c; Quick changeover, or SMED, mitigates the hidden costs of producing and transporting large amounts of inventory. [V.C.2]

41. a; The following are items to consider when focusing on improving internal setup times: standardized tools, fasteners, and placement. [V.C.2]

42. a; To reduce changeover times, the aim is to move activities currently performed when the machine is shut down to either before or after the shutdown. This is known as transforming internal work to external work. [V.C.2]

43. c; Some of the major activities included in the guidelines are setting the target, selecting the leaders, checking the operating lines, and confirming the inventory. [V.C.3]

44. b; In a kaizen blitz, the pace of improvements is more aggressive than with kaizen. [V.C.3]

Section VI

Control Phase

(Exam Targets 15 Questions)

A. SPC

1. SPC Basics

Describe the theory and objectives of SPC, including measuring and monitoring process performance for both continuous and discrete data. Define and distinguish between common and special cause variation and how these conditions can be deduced from control chart analysis. (Analyze)

2. Rational subgrouping

Define and describe how rational subgrouping is used. (Understand)

3. Control charts

Identify, select, construct, and use control charts: $\bar{X} - R$, $\bar{X} - s$, individual and moving range (ImR or XmR), median, p, np, c, and u. (Apply)

B. Sustain improvements

1. Control plan

Assist in developing and implementing a control plan to document and monitor the process. (Apply)

2. Document control

Understand document control and its role in controlling and sustaining improvements. (Understand)

3. Training plans

Develop training plans to implement and sustain improvements. (Apply)

4. Audits

Define first-, second-, and third-party audits. (Remember)

5. Plan-do-check-act (PDCA)

Apply and distinguish between the steps of plan-do-check-act (PDCA). (Apply)

C. Lean tools for process control

1. Total productive maintenance (TPM)

Define the elements of TPM, including use of predictive maintenance and describe how they can be used to control the improved process. (Understand)

2. Visual factory

Define the elements of a visual factory (andon, jidoka) and describe how they can be used to control the improved process. (Understand)

QUESTIONS

1. The observed change caused by the interplay of multiple minor variables in the process is known as a/an:

   a. chance or unassignable cause.

   b. assignable cause.

   c. statistical process display.

   d. null hypothesis.

2. Tampering or impulsively modifying the process to reduce variation will:

   a. ultimately have no effect on the process.

   b. contribute to increasing variation.

   c. correct the common and special causes of variation more quickly.

   d. reduce the outliers and reduce the process range.

3. The variation associated with the sequence of activities reflects the:

   a. voice of the control plan.

   b. voice of the enterprise.

   c. voice of the customer.

   d. voice of the process.

4. Unusual events that the operator can remove or adjust are known as:

   a. special causes of variation.

   b. common causes of variation.

c. natural variation.

d. unassignable causes of variation.

5. Which statement is true concerning the relationship between data types and control charts?

   a. Control charts for variables data require measurement on discrete either/or charts.

   b. Control charts for attributes data require measurement on a continuous scale.

   c. Control charts for variables data require measurement on a continuous scale.

   d. Control charts for attributes data require measurement on both discrete either/or charts and continuous scales.

6. Which of the following distributions can be used to model continuous data?

   a. Hypergeometric

   b. Binomial

   c. Poisson

   d. Normal

7. When choosing subgroups for rational subgrouping, one objective is to:

   a. minimize piece-to-piece variation within the subgroup.

   b. have different sample sizes for all subgroups.

   c. shield from recently collected data to avoid intrusions.

   d. make use of kanban and takt times.

8. Pieces from five spot-welding stations flow onto a common conveyor belt. An operator chooses five consecutive pieces each hour to test the tensile strength of the welds. How might this procedure change once an X-bar R chart is used to track the tensile strength?

   a. Rational subgrouping will have to be considered.

   b. The tensile strength must be converted to pass/fail data.

   c. The specifications for tensile strength will have to be adjusted.

   d. No changes will be needed.

9. When quality is measured in counts, fractions, or percentages of defective items, the appropriate control chart to select is a(n):

   a. variables control chart.

   b. attributes control chart.

   c. individual and moving range control chart.

   d. median control chart.

10. A control chart will be created to monitor a process with an automated vision system to measure height, the critical feature. A subgroup size of 15 will be used. Which of the following control charts is best suited for this application?

    a. *XmR* chart

    b. X-bar and *S* chart

    c. *np* chart

    d. c-chart

11. A powder coating line will be monitored for quality. The number of defects per item will be counted using a constant sample size. Which of the following control charts is best suited for this application?

    a. *p*

    b. *np*

    c. *XmR*

    d. *c*

12. An *R* chart is used rather than an *S* chart because it is:

    a. easier to hand-calculate.

    b. more sensitive to variation.

    c. more accurate.

    d. useful for large subgroup sizes.

13. Calculate the upper control limit of an X-bar chart using the following information:

$$\bar{\bar{X}} = 90.475,\ n = 6,\ \bar{R} = 12.440$$

    a. 84.4665

    b. 96.4835

c. 97.6529

d. 108.2270

Use the following information to answer questions 14 and 15.

$$\bar{\bar{X}} = 29.87, \bar{s} = 3.55, n = 16$$

14. Calculate the upper and lower control limits for the *S* chart.

    a. [1.232, 5.868]

    b. [1.289, 5.811]

    c. [1.590, 5.510]

    d. [1.519, 5.581]

15. Calculate the upper and lower control limits for the X-bar chart.

    a. [27.16, 32.58]

    b. [29.11, 29.93]

    c. [29.19, 30.55]

    d. [29.70, 30.04]

Use the following information to answer questions 16 and 17.

A single sample was measured each hour over the course of 25 hours. An individual moving range (ImR) chart with period n = 2 will be constructed using the following summary statistics:

$$\sum_{i=1}^{k} X_i = 674.50, \sum_{i=1}^{k} mR_i = 24.40$$

16. What are the control limits of the individual chart?

    a. [0, 36.29]

    b. [24.05, 29.91]

    c. [24.27, 29.69]

    d. [26.20, 27.76]

17. What are the control limits of the moving range chart?

    a. [0, 2.71]

    b. [0, 3.32]

c. [0.47, 1.57]

d. [0.57, 1.46]

18. A *p*-chart is used for:

a. monitoring the count of nonconformities in a product.

b. monitoring the fraction nonconforming.

c. cases in which there are few opportunities for nonconformities to occur.

d. continuous data.

19. What action should be taken if the lower limit of a *c*-chart is negative?

a. Plot as calculated.

b. Set the lower limit to zero.

c. Recalculate. An error has been made.

d. Use a *u*-chart.

20. A *u*-chart measures the:

a. count of nonconformities.

b. rate of defectives per sample.

c. count of defectives per unit.

d. rate of nonconformities per unit.

21. A *c* control chart would be best to use in which of the following situations?

a. The number of defects per washing machine produced will be tracked.

b. The number of nonconforming light bulbs in a constant sample size of 300 bulbs will be tracked.

c. The fraction of defective printed circuit boards produced each shift will be tracked. The number of boards produced per shift can vary from 400 to 500, depending on the type of chip being manufactured.

d. The average number of flaws per computer produced will be tracked. Random samples of 20–30 computers are checked each hour.

22. An *np*-chart is used for monitoring:

a. the number of nonconformities in a product.

b. the fraction nonconforming in a lot.

c. cases where there are few opportunities for a nonconformity to occur.

d. the number of nonconforming products.

Use the following information to answer questions 23 and 24.

$$\sum np = 55, \sum n = 2800, k = 30$$

23. What is the centerline of the p-chart?

a. 0

b. 0.0007

c. 0.0196

d. 0.5893

24. Calculate the control limits of the p-chart using the average sample size.

a. [0, 0.0086]

b. [0, 0.0627]

c. [0.0118, 0.0275]

d. [0.4540, 0.6570]

Use the following information to answer questions 25 and 26.

$$\sum np = 78, k = 25, n = 200$$

25. What is the centerline of the np-chart?

a. 0.016

b. 0.390

c. 3.120

d. 5.791

26. What are the upper and lower control limits of the np-chart?

a. [0, 1.30]

b. [0, 7.26]

c. [0, 8.38]

d. [2.46, 3.78]

Use the following information to answer the questions 27 and 28.

$$\sum_{i=1}^{k} c_i = 240, \sum_{i=1}^{k} n_i = 1550, k = 25$$

27. What is the centerline of the u-chart?
    a. 0.006
    b. 0.154
    c. 3.85
    d. 9.60
28. What are the upper and lower control limits for a point with $n = 20$ samples?
    a. [0, 0.059]
    b. [0.131, 0.177]
    c. [0, 0.417]
    d. [0.1241, 0.1838]
29. The control plan explains how to control:
    a. the nonconforming products or out-of-control conditions.
    b. the design of experiments.
    c. the safety characteristics in the design.
    d. the workflow in your process.
30. Controls can be visually represented by:
    a. gages, sample sizes, and frequency of testing.
    b. display of special characteristics.
    c. flowcharts and graphical symbols.
    d. confirmation by stakeholders of the process under control.
31. Visual inspections of process conditions, first pieces produced, and incoming materials are effective ways to:
    a. reveal failure modes early before expensive work is performed.
    b. demonstrate the validity of sampling plans.
    c. record and document initial process outcomes.
    d. prevent the process controls from becoming obsolete or irrelevant.
32. The primary purpose of the control plan is to:
    a. implement validations on specific processes.
    b. ensure that process changes are maintained over time.

c. create a structure for implementing quality throughout the process.

d. keep only one item as a variable and maintain all other characteristics as controls per experiment.

33. Developing documents such as SOPs, work instructions, and control plans helps ensure that:

a. the paperwork burden remains high.

b. internal auditors are kept busy.

c. improvements are sustained over time.

d. the right people get credit for the improvement.

34. Which of the following tools is used to monitor a process to ensure that the process improvements are continuing to produce the desired results?

a. SOPs

b. Control charts

c. Gage R&R

d. Control plan

35. Standard operating procedures (SOPs) and work instructions can:

a. reduce process variation.

b. allow for gradual improvement over time.

c. aid in benchmarking efforts.

d. create an atmosphere of mistrust.

36. The dynamic control plan is a living document whereby:

a. document control practices are suspended during operations.

b. standard operating procedures, failure modes, and effects analysis contradict each other.

c. root causes and corrective actions are applied to future versions.

d. operators have the right and responsibility to make updates any time that things change.

37. Dynamic control planning supports control plans with:

a. pictures, colors, and shapes.

b. spreadsheets filled with numbers.

c. illustrations, instructions, and procedures for implementation and maintenance.

d. work instructions and user manuals.

38. How should the quality functional leadership respond if operators propose adjustments to documents?

a. Reprimand the operators, as adjustments are not permitted.

b. Discard the documents, as the need for adjustments invalidates their suitability.

c. Review the proposed adjustments to remove nonessential portions while inserting necessary components into new versions.

d. Freely permit all operators to adjust documents at will.

39. What is necessary to ensure coverage of materials and sustain improvements through targeted propagation of new methods and practices to the affected participants and stakeholders?

a. FMEA

b. Training plans

c. Control plans

d. Profound knowledge

40. What is the first step in developing training plans?

a. Needs analysis and business case

b. Strategic planning

c. Definition of learning experiences

d. Training documentation, facilitation, and provision

41. What type of supporting activities and devices support successful process completion in the control phase?

a. Consensus decisions

b. Training

c. Requirements elicitation

d. Organic environmental materials

42. Training is distinguished from general education by the expectation that the outcome of successful training will result in:

    a. specific competencies.

    b. an increased attention span.

    c. improved test taking.

    d. improved teamwork.

43. What should training objectives be expressed as?

    a. Needs analysis

    b. Business cases

    c. Cost-and-benefit relationships

    d. Performance targets

44. When an audit is conducted by local personnel on their own property, it is considered what type of audit?

    a. First-party audit

    b. Second-party audit

    c. Third-party audit

    d. Fourth-party audit

45. When an audit is conducted by a registrar in your organization, the audit type is called a:

    a. first-party audit.

    b. second-party audit.

    c. third-party audit.

    d. fourth-party audit.

46. A key factor in using ISO 19011 audits in Six Sigma is to:

    a. ensure that the root cause of the problem was identified correctly.

    b. sustain improvements by reinforcing those practices that contributed to improvement.

    c. verify that the operator work instructions are being followed as designed by the process.

    d. verify management commitment to the continual improvement of the operations.

47. A type of quality assessment activity intended to reconcile the description of systems, processes, activities, and objective evidence with their actual outcomes is a(n):

    a. FMEA.

    b. tollgate.

    c. Gemba.

    d. audit.

48. How are third-party audits different from second-party audits?

    a. Second-party audits are conducted by external providers.

    b. Third-party audits are conducted by external providers.

    c. Third-party audits are conducted by a representative of a regulatory or accreditation body.

    d. Third-party audits are conducted by external providers as a condition of continued purchases or preservation of the business relationship.

49. The correct representation of audit conclusions, audit criteria, and audit findings is:

    a. audit findings (objective evidence and observations) are compared to audit criteria (expected references and behavior), leading to audit conclusions.

    b. audit criteria (objective evidence and observations) are compared to audit findings (expected references and behavior), leading to audit conclusions.

    c. audit conclusions (objective evidence and observations) are compared to audit findings (expected references and behavior), leading to audit criteria.

    d. audit conclusions (objective evidence and observations) are compared to audit criteria (expected references and behavior), leading to audit findings.

50. At what stage of the PCDA cycle is the process or problem selected for improvement described?

    a. Act

    b. Check/study

    c. Do

    d. Plan

51. Which option correctly shows the relationship between the phase of PDCA and DMAIC and the corresponding actions?
    a. "Act" maps to the control phase. Actions include implementing solutions and preventing recurrence.
    b. "Check" maps to the measure phase. Actions include defining the problem and interim containment.
    c. "Do" maps to the define phase. Actions include identifying the problem and using the team approach.
    d. "Plan" maps to the analyze and improve phases. Actions include defining the root cause and developing solutions.
52. TPM aspires to maximize uptime along with:
    a. timely equipment repairs after breakdown.
    b. scheduled maintenance activities.
    c. machine performance and first-pass yield.
    d. equipment engineering economics.
53. A baseline measurement of overall equipment effectiveness (OEE) is the product of:
    a. cost, quality, and speed.
    b. availability, performance efficiency, and quality rate.
    c. mean time to repair and mean time between failures.
    d. maintenance, data collection, and workflow.
54. Leading indicators of potential equipment breakdowns detected through TPM include:
    a. minor stoppages and reduced speed.
    b. industry and benchmark data.
    c. data collection and operational involvement.
    d. maintenance infrastructure and applications.
55. Operators can enhance TPM through:
    a. equipment management and prevention of contamination.
    b. planned and preventive maintenance of equipment.

c. inspection and collection of data to monitor and measure deterioration.

d. training and accountability for operating equipment.

56. Unplanned downtime loss is defined as the sum of the losses of:

a. equipment operation due to operating at less than maximum designed efficiency.

b. equipment availability due to unplanned stoppage of production over a period of time.

c. equipment runs relative to intended design or cycle times.

d. first-pass yield outcomes to planned available time.

57. Predictive maintenance is best defined by:

a. improving or modifying equipment to prevent breakdowns or make maintenance easier.

b. scheduled maintenance activities performed to prevent unforeseen breakdowns.

c. the act of repairing a piece of equipment after it breaks down.

d. replacing components that are likely to fail based on analysis.

58. Visual factory models support cross-trained personnel and mixed-model schedules by communicating:

a. process goals, operating conditions, and known problems.

b. product specifications and dimensions.

c. explicit work instructions and project charts.

d. Six Sigma improvement specifications.

59. A single-point lesson relies on visual displays for:

a. schedule boards.

b. problems and employee notifications.

c. maintenance and sustainability.

d. task performance and inspection methods.

60. Employees and management must respond to which of these active visual signals in the workplace?

    a. Alarm lights, depleted stock, visual defects, and safety displays

    b. Pictures, colors, and shapes

    c. Written procedures and work instructions

    d. Spreadsheets filled with numbers

61. Jidoka or *autonomation* integrates automation with human intelligence to provide which of the following advantages?

    a. Reduces the need for direct monitoring but enables responses based on human judgment

    b. Reduces the need for headcount, enabling right-sizing of surplus factory staff

    c. Reduces common cause variation

    d. Increases the need for direct monitoring and multi-process handling

62. At a grocery store, each checkout aisle has a light indicating whether it is open or closed, or if the cashier is waiting for price information. This is an example of:

    a. just in time.

    b. andon.

    c. kaizen.

    d. predictive maintenance.

SOLUTIONS

1. a; Chance or unassignable cause occurs when the observed change is caused by the interplay of multiple minor variables in the process. [VI.A.1]
2. b; Tampering will contribute to increasing variation. [VI.A.1]
3. d; Process variation reflects the voice of the process. [VI.A.1]
4. a; Special causes of variation include unusual events. [VI.A.1]
5. c; Control charts for variables data require measurement on a continuous scale. [VI.A.1]
6. d; The normal distribution is a probability distribution that can be used to model continuous data. [VI.A.1]
7. a; Choose subgroups to minimize piece-to-piece variation. [VI.A.2]
8. a; Rational sub-grouping will have to be considered. Now, samples are a mixture of pieces from five different stations. [VI.A.2]
9. b; Attributes control charts require only a count of defective parts or services rather than precise and careful measurements. [VI.A.3]
10. b; Use an X-bar and *S* chart when measuring continuous variables with subgroup sizes greater than or equal to 10. [VI.A.3]
11. d; Use a *c*-chart when counting defects per sampling unit with a constant sample size. [VI.A3]
12. a; An *R* chart is used rather than an *S* chart because it is easier to hand-calculate. [VI.A.3]
13. b; To calculate the upper control limit of the X-bar chart, we need to look up the table value A_2, which is based on the subgroup size, *n*. Then we use:

$$UCL_{\bar{X}} = \bar{\bar{X}} + A_2\bar{R}$$
$$\text{A}_2 \text{ for n} = 6 \text{ is } 0.483$$

$$UCL_{\bar{X}} = 90.475 + 0.483 \times 12.440 = 96.4835.$$

[VI.A.3]

14. c; To calculate an S chart's upper and lower control limits, we need to look up two table values, B_3 and B_4, based on the subgroup size, n. Then we use:

$$LCL_S = B_3\overline{S}$$
$$B_3 \text{ for } n = 16 \text{ is } 0.448$$

$$LCL_S = 0.448 \times 3.55 = 1.590$$

$$UCL_S = B_4\overline{S}$$
$$B_4 \text{ for } n = 16 \text{ is } 1.552$$

$$UCL_S = 1.552 \times 3.55 = 5.510.$$

[VI.A.3]

15. a; To calculate the control limits of the X-bar chart with $n = 16$, we find the table value $A_3 = 0.763$ based on n. Then

$$LCL = \overline{\overline{X}} - A_3\overline{s} = 29.87 - 0.763 \times 3.55 = 27.16$$
$$LCL = \overline{\overline{X}} + A_3\overline{s} = 29.87 + 0.763 \times 3.55 = 32.58.$$

[VI.A.3]

16. c; We note that the number of samples is $k = 25$ and the period for the moving average is $n = 2$. We first calculate

$$\overline{X} = \frac{\sum_{i=1}^{k} X_i}{k} = \frac{674.5}{25} = 29.98$$

$$\overline{mR} = \frac{\sum_{i=1}^{k} mR_i}{k - 1} = \frac{24.40}{24} = 1.017.$$

We find the table value E_2 for $n = 2$ to be 2.660. Then

$$LCL = \overline{X} - E_2\overline{mR} = 26.98 - 2.660 \times 1.017 = 24.27$$
$$UCL = \overline{X} + E_2\overline{mR} = 26.98 + 2.660 \times 1.017 = 29.69.$$

[VI.A.3]

17. b; To find the control limits for the moving range chart with period $n=2$, we look up table values $D_3=0$ and $D_4=3.267$ based on n. Then

$$LCL = D_3\overline{mR} = 0 \times 1.017 = 0$$
$$UCL = D_4\overline{mR} = 3.267 \times 1.017 = 3.32.$$

[VI.A.3]

18. b; A p-chart is used to monitor the fraction nonconforming. [VI.A.3]
19. b; If the lower limit of a c-chart is negative, set the lower limit to zero. [VI.A.3]
20. d; A u-chart measures the rate of nonconformities (defects) per unit. [VI.A.3]
21. a; The c-chart would be best to track the number of surface defects on stainless steel panels in lots of constant sample size. [VI.A.3]
22. d; An np-chart is used to monitor the number of nonconforming products. The np-chart uses a constant sample size. [VI.A.3]
23. c; To find the average proportion defective, we use

$$\overline{p} = \frac{\sum_{i=1}^{k} D_i}{\sum_{i=1}^{k} n_i} = \frac{\sum_{i=1}^{k} n_i p_i}{\sum_{i=1}^{k} n_i} = \frac{55}{2800} = 0.0196.$$

[VI.A.3]

24. b; To find the average n, we use

$$\overline{n} = \frac{\sum_{i=1}^{k} n_i}{k} = \frac{2800}{30} = 93.3.$$

Then, we can calculate the control limits of the p-chart:

$$LCL = \overline{p} - 3\sqrt{\frac{\overline{p}(1-\overline{p})}{\overline{n}}} = 0.0196 - 3\sqrt{\frac{0.0196(1-0.0196)}{93.3}} = -\,0.0234 \uparrow 0$$

$$LCL = \overline{p} + 3\sqrt{\frac{\overline{p}(1-\overline{p})}{\overline{n}}} = 0.0196 + 3\sqrt{\frac{0.0196(1-0.0196)}{93.3}} = 0.0627.$$

[VI.A.3]

25. c; To find the centerline of the *np*-chart, we use

$$\overline{np} = \frac{\sum_{i=1}^{k}(np_i)}{k} = \frac{78}{25} = 3.120.$$

[VI.A.3]

26. c; To calculate the control limits of the *np*-chart, we first must find *p*-bar:

$$\bar{p} = \frac{\sum_{i=1}^{k}(np_i)}{kn} = \frac{78}{25 \times 200} = 0.0156.$$

Then,

$$LCL = n\bar{p} - 3\sqrt{n\bar{p} \times (1-\bar{p})} = 3.12 - 3\sqrt{3.12(1-0.0156)} = -2.138 \uparrow 0$$

$$LCL = n\bar{p} + 3\sqrt{n\bar{p} \times (1-\bar{p})} = 3.12 + 3\sqrt{3.12(1-0.0156)} = 8.378.$$

[VI.A.3]

27. b; To calculate the centerline for the *u*-chart, we use

$$\bar{u} = \frac{\sum_{i=1}^{k} c_i}{\sum_{i=1}^{k} n_i} = \frac{240}{1550} = 0.1539.$$

The variable *k* is the number of subgroups used for the control limits. In this example, 240 defects were found in 25 subgroups, and 1550 units were inspected. [VI.A.3]

28. c; To calculate the control limits for the *u*-chart, we note that u-bar = 0.154, as shown in the previous question, and $n_i = 20$. Then we use

$$\bar{u} = 0.154 \; and \; n_i = 20$$

$$LCL = \bar{u} - 3\sqrt{\frac{\bar{u}}{n_i}} = 0.154 - 3\sqrt{\frac{0.154}{20}} = -0.109 \uparrow 0$$

$$LCL = \bar{u} + 3\sqrt{\frac{\bar{u}}{n_i}} = 0.154 + 3\sqrt{\frac{0.154}{20}} = 0.417.$$

[VI.A.3]

29. d; The control plan is a document explaining how to control your process's workflow. [VI.B.1]
30. c; Control plans include flowcharts and graphical symbols for visual representation. [VI.B.1]
31. a; Visual inspections reveal failure modes early—before expensive work is performed. [VI.B.1]
32. b; The primary purpose of a control plan is to ensure that process changes are maintained over time. [VI.B.1]
33. c; Developing documents such as SOPs, work instructions, and control plans help ensure that improvements are sustained over time. [VI.B.1]
34. b; Control charts are used to monitor a process and make sure that improvements continue to produce the desired results. [VI.B.1]
35. a; SOPs and work instructions help ensure that procedures are performed the same way each time, which reduces process variation. [VI.B.1]
36. d; Operators have the right and responsibility to update the dynamic control plan. [VI.B.2]
37. c; Illustrations, instructions, and procedures support control plans. [VI.B.2]
38. c; If operators need to make adjustments, these should be reviewed to remove nonessential portions while inserting necessary components into new versions. [VI.B.2]
39. b; Training plans ensure that affected participants and stakeholders receive coverage of materials on new methods and practices. [VI.B.3]
40. a; The first step in developing training plans is to determine the needs analysis and business case to justify the training. [VI.B.3]
41. b; Training supports the successful completion of the process in the control phase by demonstrating specific competencies. [VI.B.3]
42. a; Training is distinguished from general education by the expectation of specific competencies in the subject area or task domain being trained. [VI.B.3]
43. d; Training objectives should be expressed as performance targets. There should be a clear mapping between the training activities and the desired performance. [VI.B.3]
44. a; First-party audit (also described in ISO 19011:2018). An audit conducted by local personnel on their own property is considered a first-party audit. [VI.B.4]
45. c; Third-party audit (also described in ISO 19011:2018). An audit conducted by a registrar on your organization is called a third-party audit. [VI.B.4]

46. b; In Six Sigma, a key factor of using ISO 19011 audits is to sustain improvements. This is done by reinforcing those practices that contributed to the improvement. [VI.B.4]

47. d; An audit is a type of quality assessment activity intended to reconcile the description of systems, processes, activities, and objective evidence with their actual outcomes. [VI.B.4]

48. c; Third-party audits are conducted by a representative of a regulatory or accreditation body. [VI.B.4]

49. a; Audit findings (objective evidence and observations) are compared to audit criteria (expected references and behavior), leading to audit conclusions. [VI.B.4]

50. d; At the *plan* stage, the process or problem that is selected for improvement is described, and the improvement goals and objectives are defined. [VI.B.5]

51. a; The PDCA act stage is mapped to control, prioritizing implementing the solutions and preventing recurrence. [VI.B.5]

52. c; TPM also seeks to improve machine performance and first-pass yield. [VI.C.1]

53. b; The equation for OEE is the product of availability, performance efficiency, and quality rate. [VI.C.1]

54. a; Early signs of a breakdown include stoppages and speed reductions, the detection of which can prevent breakdowns and improve understanding. [VI.C.1]

55. c; Preventing, correcting, and measuring deterioration will enhance TPM. [VI.C.1]

56. b; Unplanned downtime loss is the sum of losses due to unplanned stoppage of production. [VI.C.1]

57. d; Predictive maintenance activities include proactively replacing components that are likely to fail based on analysis. [VI.C.1]

58. a; The visual factory strives to make problems visible, notify employees of current operating conditions, and communicate process goals. [VI.C.2]

59. d; A single-point lesson shows task performance and inspection of a product or service. [VI.C.2]

60. a; Visual signals include alarm lights, depleted stock, visual defects, and safety displays that warn against entry into a danger zone without protection. [VI.C.2]

61. a; The integration of automation with human intelligence reduces the need for direct monitoring but enables responses based on human judgment. [VI.C.2]

62. b; Having visual signals such as lights that show the status of machines or processes is known as andon. [VI.C.2]

Part B

Practice Exam

QUESTIONS

1. Who is the quality guru who created the 14 Points and the 7 Deadly Diseases?
   a. W. Edwards Deming
   b. Joseph M. Juran
   c. Kaoru Ishikawa
   d. Walter A. Shewhart
2. When an individual or team develops as many ideas concerning a topic as they can in a short period of time using various creativity techniques, it is called:
   a. strategic planning.
   b. brainstorming.
   c. affinity technique.
   d. multi-voting.
3. Which of the following elements demonstrates the outcomes of jidoka or autonomation?
   a. Prevention of stoppages and responses to quality issues and irregularities and detection of common cause variation
   b. Automated stoppages and responses to quality issues and irregularities and detection of special cause variation
   c. Mitigation of stoppages and responses to quality issues and irregularities and detection of zero-defect variation
   d. Increases in direct monitoring and manual handling
4. A *p*-chart would be best to use in which of the following situations?
   a. Tracking the number of surface defects in lots of 10 stainless steel panels
   b. Tracking the number of nonconforming light bulbs in a constant sample size of 300 bulbs
   c. Tracking the fraction of defective printed circuit boards produced each shift when the number of boards produced per shift can vary from 400 to 500, depending on what type of chip is being manufactured
   d. Tracking the average number of flaws per computer produced when random samples of 20–30 computers need to be checked each hour

5. The purpose of a RACI matrix is to:

   a. define the severity of failure modes found in the risk base tool FMEA.

   b. describe and define the roles and responsibilities of team members.

   c. identify the relationships between product features.

   d. record the outcome of a 5 Whys exercise.

6. The term kaizen refers to:

   a. mistake-proofing.

   b. constraint reduction.

   c. continuous improvement.

   d. signals or lights.

7. What action can we take on a null hypothesis?

   a. Reject or fail to reject

   b. Reject or accept

   c. Accept

   d. Accept or fail to reject

8. Which quality tool uses pre-categorized potential outcomes for data collection using sets of words, tally lists, or graphics?

   a. Scatter diagrams

   b. Run charts

   c. Pareto diagrams

   d. Check sheets

9. Which chart helps teams narrow their focus based on the 80/20 rule?

   a. Histogram

   b. Prioritization

   c. Pareto

   d. Affinity

10. Which graphic tool is best for developing a complete picture of the relationship between two variables?

    a. Histogram

    b. Pareto chart

c. Cause-and-effect diagram

d. Scatter plot

11. Process performance is a measure of how the process is executed against an established goal. The process performance measures usually include:

a. defects per time period, defects per opportunity, mean time between failures, and mean time to repair.

b. defects in a process mean, defects in the capacity ratio, and sigma levels.

c. rolled throughput yield, a mean divided by sigma, and defects per time period.

d. defects per unit, parts per million, rolled throughput yield, sigma levels, and process capability indices.

12. The likelihood that current controls will prevent a failure from reaching the customer is called:

a. detection.

b. occurrence.

c. severity.

d. FMEA.

13. What does the following symbol on a value stream map represent?

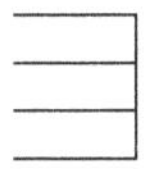

a. Safety or buffer stock

b. Production kanban

c. Inventory

d. Supermarket

14. Cause-and-effect diagrams can be used to analyze:

a. all suspicious outlying effects.

b. factors in a production cycle.

c. the use of endcaps in marketing.

d. positive effects as well as undesirable ones.

15. 5S is one of the first tools to apply in the path to achieving lean enterprise organizations. The traditional sequence for 5S includes which of the following?
    a. Sort, set in order, shine, standardize, and sustain
    b. Set in order, strategic catalog, shine wax parts, and sustain gains
    c. Situate in alphabetical order, select each part, and set into appropriate containers
    d. Strategize and plan, select dirty and damaged items, sanitize with organic cleansers, sterile dry, and situate into storage

16. A half-fraction of a 2^6 experiment has how many runs?
    a. 6
    b. 12
    c. 32
    d. 64

17. From which distribution does a confidence interval on the variance of a normal distribution use critical values?
    a. Chi-square
    b. Poisson
    c. Binomial
    d. Standard normal

18. Find the critical value for an F statistic with 3 and 30 degrees of freedom and $\alpha = 0.01$.
    a. 2.92
    b. 4.51
    c. 5.42
    d. 26.50

19. Mistake-proofing is achieved by:
    a. training and reinforcement of proper work instructions.
    b. limiting or restricting ways to complete a task.
    c. adherence to process control charts.
    d. constant supervision and rapid intervention.

20. When a team is in the performing growth stage, team members:

    a. get to know each other.

    b. voice their ideas.

    c. resolve their conflicts.

    d. realize their interdependence.

21. For quick changeovers, setup time is defined as the time from the last piece of one product to:

    a. the completion of the machine changes.

    b. the first good piece of another product.

    c. the last piece of the other product.

    d. the next changeover cycle.

22. The one-way ANOVA table includes which sources of variation?

    a. Confidence level, significance level

    b. Mean squares, F-statistic

    c. Readings, levels, measurements

    d. Between treatment, within treatment, total

23. Which graphical tool provides a pictorial view of numerical data about location and spread?

    a. Stem-and-leaf

    b. Scatter diagram

    c. Pareto chart

    d. Run chart

24. A Green Belt performs a chi-square goodness-of-fit test with alpha = 0.05 to determine whether the distribution of defect types last month differed from the historical percentages. Given the data table below, which statement is true?

	Defect Type		
	Length	Weight	Strength
Historical %	50%	25%	25%
Monthly inspection results	60	35	45

    a. The test statistic = 3.889, and the critical value = 5.991.

    b. The test statistic = 4.286, and the critical value = 5.991.

c. The test statistic = 3.889, and the critical value = 7.815.

d. The test statistic = 4.286, and the critical value = 7.815.

25. Which of the following subgroup sizes for an $\bar{X}$ and R chart would most likely be used if measurement were costly?

    a. 1

    b. 2

    c. 3

    d. 4

26. Warranty data, complaint logs, and focus groups are inputs for:

    a. voice of the customer.

    b. value stream mapping.

    c. kaizen blitz.

    d. process FMEA.

27. A sample of size five is randomly selected from a process with 15 percent defective. Find the probability that the sample has exactly one defective.

    a. 0.3685

    b. 0.3915

    c. 0.3993

    d. 0.4096

28. Risk management can be defined using the PDCA process. Which item is part of *Plan*?

    a. Implementing a test of the chosen risk management technique(s)

    b. Monitoring the implemented techniques for adequacy of protection

    c. Assessing the effectiveness of the technique and making necessary adjustments

    d. Identifying and defining potential exposures to loss

29. Rank the rows in the FMEA table from largest to smallest RPN.

Row	Severity	Occurrence	Detection
1	5	5	5
2	8	5	2
3	7	4	4
4	3	6	5

a. 1, 2, 3, 4

b. 1, 3, 4, 2

c. 1, 2, 4, 3

d. 1, 4, 3, 2

30. When conducting an MSA study, a common error made while performing GR&R is:

    a. altering the samples during the study.

    b. using parts that cover the range of the gage.

    c. using overtrained inspectors.

    d. using parts that cover the specification range.

31. Variables that show the observed results of an experimental treatment as a result of controlling levels, interactions, and a number of other factors are:

    a. dependent variables.

    b. response variables.

    c. independent variables.

    d. main effects.

32. How are training objectives proven to be successfully adopted and demonstrated?

    a. Survey from the training attendees with positive commentary

    b. Objective evidence of successful work from process changes and improvements

    c. Completion of third-party certifications in quality and Six Sigma management

    d. Subscription to online programs and webinars for continuity of additional training

33. In a multi-vari chart study, another term for *part-to-part* or *lot-to-lot* variation is:

    a. positional variation.

    b. cyclical variation.

    c. normal variation.

    d. temporal variation.

34. The weight of soup cans follows a normal distribution with a mean of 16.0 oz and a standard deviation of 0.005 oz. What percentage of cans weighs more than 16.012 oz?

    a. 0.82%

    b. 40.52%

    c. 59.48%

    d. 99.18%

35. A visual workplace supports visual devices that are easily understood and accessed by all through the use of:

    a. spreadsheets filled with numbers.

    b. pictures, colors, shapes.

    c. written procedures.

    d. instructions from supervisors.

Use the ANOVA table below to answer questions 36 and 37.

Source	SS	DF	MS	F
Machines	456.7	4		
Error	123.8	24		
Total	580.5	28		

36. What is the value of the F statistic in the table?

    a. 0.61

    b. 0.79

    c. 3.69

    d. 22.13

37. If the test were conducted at an alpha level of 0.05, which of the following statements is the correct conclusion?

    a. $F_{Calc} > F(24, 4, 0.05)$, Reject H_0.

    b. $F_{Calc} < F(4, 24, 0.95)$, Accept H_0.

    c. $F_{Calc} > F(4, 24, 0.05)$, Reject H_0.

    d. $F_{Calc} < F(4, 24, 0.05)$, Accept H_0.

38. Calculate the lower control limit of an R chart using the following information:

$$\bar{\bar{X}} = 345.50,\ n = 7,\ \bar{R} = 16.25$$

a. 0

b. 1.235

c. 9.441

d. 12.350

39. A quality inspector tallies the number of paint defects on 25 car door panels. Which probability distribution is associated with these defect counts?

a. Normal

b. Binomial

c. F

d. Poisson

40. The purpose of randomizing the order of experimental runs is to:

a. make the experiment harder to conduct.

b. separate the runs into blocks.

c. reduce interactions of the factors.

d. minimize the effect of uncontrolled variables.

41. Process inputs for a fishbone diagram are traditionally recognized as:

a. manpower, materials, communication, weather, and executives.

b. manpower, materials, methods, Mother Nature, management, and measurement.

c. humans, time, machinery, environment, and coworkers.

d. employees, material, methods, environment, supervisors, and calibration.

42. A common way for a Green Belt team to make a team decision is through:

a. a SWOT analysis.

b. SMED.

c. multi-voting.

d. an ANOVA analysis.

43. A hypothesis test yields a p-value equal to 0.035. If the test is conducted at $\alpha = 0.05$, what is the conclusion?

    a. Accept H_0 since $p < \alpha$.

    b. Reject H_0 since $p < \alpha$.

    c. Do not reject H_0 since $2p > \alpha$.

    d. It depends on the type of test conducted.

44. Flow utilized when a line operator or front desk personnel want to provide feedback to the management, or a team member wants to communicate to the executive sponsor of the project is called:

    a. matrix flow.

    b. lateral flow.

    c. upward flow.

    d. diagonal flow.

45. Sampling from a lot in which each unit has an equal probability of getting chosen is known as:

    a. random sampling.

    b. sequential sampling.

    c. stratified sampling.

    d. biased sampling.

46. A control plan lists action steps to take if the critical parameter goes out of control. These actions would most likely be found in which column of the control plan?

    a. Reaction plan

    b. Characteristic

    c. Specification

    d. Frequency

47. Lean experts define a process step as value-added if:

    a. the manufacturer recognizes the value, it lowers the cost of the product, and it is done right the first time.

    b. exceptional design teams have been developed, the product is transformed, and it is made faster than ever before.

c. the customer recognizes the value, it changes (transforms) the product, and it is done right the first time.

d. the customer recognizes the value, the industry is globally transformed, and there are no gray areas between value and non-value.

48. This idea-generation technique does not include verbal interactions and instead involves writing ideas first and then posting them for all to read. It is:

a. force-field analysis.

b. brainstorming.

c. nominal group technique.

d. process analysis.

49. What index is calculated as the total number of defects/deficiencies divided by the total number of products produced?

a. DPO

b. DPU

c. DPMO

d. RTY

50. The specifications for the length of a new metal part are 34.0 ± 0.50 cm. Samples of size 1000 are taken on each of two shifts, resulting in mean lengths equal to 34.005 and 34.012 cm, respectively. A statistical test of the difference between mean lengths is significant at $p = 0.000$. This is an example of:

a. a paired *t* test.

b. statistical versus practical significance.

c. proactive problem-solving.

d. design of experiments.

51. A full factorial experiment:

a. always contains blocking variables.

b. tests all combinations of factor levels.

c. expresses factor levels as percentages that add to 100%.

d. has two-factor interaction confounded with main effects.

52. A value stream map (VSM) is a type of detailed process map that shows what types of information?
    a. Stakeholders, current state, and action plan
    b. WIP, changeover times, and inventory
    c. Defect locations and error types
    d. Material and worker flow and distances traveled
53. When referencing descriptive statistics, what term defines the value that is the middle of a sorted list of data?
    a. Mean
    b. Median
    c. Mode
    d. Range
54. QFD is applied to improve:
    a. the complexity of technical requirements.
    b. parts planning.
    c. customer satisfaction at an acceptable cost.
    d. the known values of a product.
55. The technique that can significantly reduce the time taken to switch from producing one product to producing another is called:
    a. countermeasure activities.
    b. quick changeover.
    c. the theory of constraints.
    d. mistake-proofing.
56. The correlation coefficient between two quantitative variables is equal to −0.75. This means that:
    a. as x decreases, y tends to decrease.
    b. as x decreases, y will be negative.
    c. as x increases, y tends to decrease.
    d. as x increases, y tends to increase.

57. Specification limits should reflect the:
    a. voice of the control plan.
    b. voice of the executive sponsor.
    c. voice of the process.
    d. voice of the customer.

58. A simple linear regression model was estimated to predict the average number of customer returns based on the number of sales the previous month:

$$\hat{y} = 230 + 0.0034x$$

    What is the expected number of returns if last month's sales equaled 45,678 units?
    a. 155.3
    b. 230.0
    c. 385.3
    d. 456.8

59. Business continuity is promoted through which ISO management system standard?
    a. ISO 22301
    b. ISO 9001
    c. ISO 14001
    d. ISO 50001

60. A spaghetti diagram is defined as a:
    a. visual representation using a continuous flow line tracing the path of an item or activity through a process.
    b. step-by-step approach for identifying all possible failures in a design, a manufacturing or assembly process, or a product or service.
    c. method that either makes it impossible for an error to occur or makes the error immediately obvious once it has occurred.
    d. picture of the separate steps of a process in sequential order, including materials or services entering or leaving the process (inputs and outputs).

61. Each hour, 10 units are sampled and inspected for defects. After 25 time periods, the total number of defects equals 38. Using this information, calculate the control limits for a *c*-chart.

    a. [0, 1.32]

    b. [0, 2.69]

    c. [0, 5.22]

    d. [0, 9.65]

62. Gage "R&R" stands for:

    a. repeatability and reproducibility.

    b. reliability and reproducibility.

    c. repeatability and reliability.

    d. relevance and restorability.

63. A project charter can take many formats but is usually limited to:

    a. an extensive assessment and analysis of the project.

    b. one or two pages serving as a top-level quick overview of the project.

    c. a project reference booklet.

    d. charts and diagrams.

64. A quality engineer analyzes five months of data taken on a process output. The average is equal to 45.67, and the standard deviation is 0.75. Specifications are 47.0 ± 3.0. What is the estimated P_p?

    a. 0.44

    b. 0.75

    c. 1.33

    d. 1.50

65. A two-factor experiment was conducted. The analysis showed that the interaction effect was not significant. Based on the main effects plots below, how will the mean response be affected if Factor A is changed from the low level to the high level?

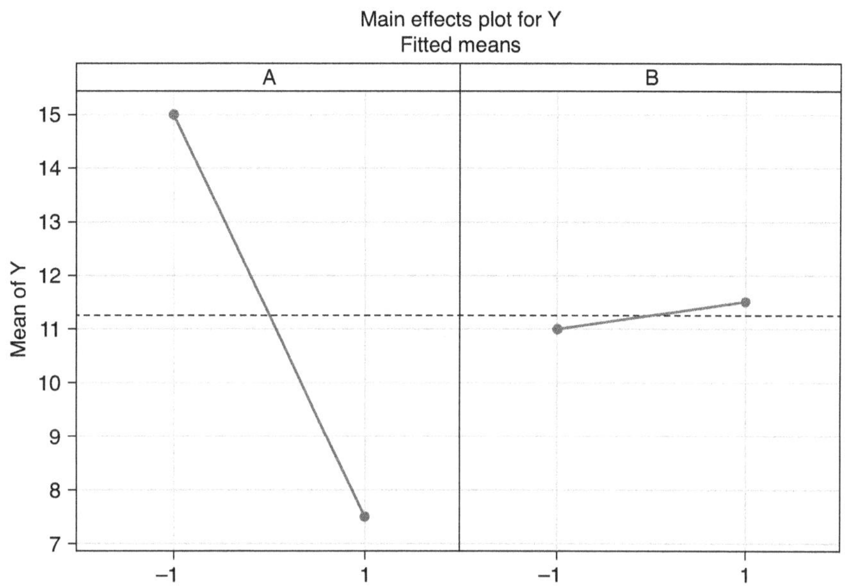

a. The mean response will decrease by 7.5 units.

b. The mean response will increase by 3.25 units.

c. The mean response will increase by 7.5 units.

d. The mean response will be dependent on the setting for Factor B.

66. The roof of a house of quality diagram displays:

a. customer requirements.

b. comparisons with the competition.

c. relationships between the technical requirements.

d. relationships between customer and technical requirements.

67. Samples of size n are taken from a symmetric distribution with a variance equal to 25. What is the variance of the distribution of sample averages?

a. $\frac{5}{\sqrt{n}}$

b. $\frac{5}{n}$

c. $\frac{25}{\sqrt{n}}$

d. $\frac{25}{n}$

68. A Green Belt runs an ANOVA test with an alpha equal to 0.01. The p value for the test is 0.002. Which of the following statements is the correct conclusion?

    a. All means are equal.

    b. All means are different.

    c. At least one mean is different from the rest.

    d. At least one variance is different from the rest.

69. Restricting ways to complete a task to ensure accuracy and compliance is called:

    a. waste.

    b. bottlenecks.

    c. poka-yoke.

    d. kanban.

70. Setting the boundaries of a project will help the team identify:

    a. solutions in the define phase.

    b. internal versus external customers.

    c. a project champion.

    d. root causes of the problem.

71. The composition and number of people on a team should be:

    a. limited to five to nine people.

    b. based on the size and scope of the project.

    c. exclude the subject matter expert and manager.

    d. based on who can actually attend the meetings in person.

72. Which document controls are used to demonstrate traceability?

    a. A description of the direct relationship between process documents and the parent references

    b. An indicator of the purpose of the document, its intended distribution, and any restrictions (i.e., internal, confidential)

c. A statement warning that printed versions are not controlled

d. A reference to the online version for the most current revision

73. Which type of data is used to count the occurrences of an event?

a. Continuous

b. Variable

c. Locational

d. Discrete

74. When information technology is applied to adjust or modify how a company receives, processes, and manages information, which technique is being applied?

a. Proof of concept

b. Try-storming simulations

c. Pilot test

d. Advanced product quality planning

75. A factor that is not of experimental interest but is included in the analysis in order to reduce experimental error is known as a(n):

a. confounding variable.

b. insignificant variable.

c. response variable.

d. blocking variable.

76. SDCA stands for:

a. Standardize, Do, Control, Adjust.

b. Systematize, Do, Control, Apply.

c. Standardize, Define, Check, Apply.

d. Standardize, Do, Check, Adjust.

77. A team succeeds in speeding up a process. What might be an unintended negative consequence?

a. Rework increases.

b. Throughput increases.

c. Defects decrease.

d. Costs decrease.

78. Which tool is used for waste elimination?

    a. Standard work

    b. Control charts

    c. Process capability

    d. Pull systems

79. What is the formula for the probability of two events occurring when the events are independent?

    a. $P(A \text{ or } B) = P(A) + P(B)$

    b. $P(A \text{ and } B) = P(A) \times P(B)$

    c. $P(A \text{ and } B) = P(A) + P(B)$

    d. $P(A \text{ or } B) = P(A) + P(B) - P(A \cap B)$

80. Which statement describes how SMART goals are used in a Six Sigma project?

    a. They push the team to achieve stretch goals.

    b. They set up five-year project targets for after completion of the DMAIC process.

    c. They are written in the control phase of a project.

    d. They help prevent scope creep across the DMAIC process.

81. Verification and validation are used to compare results against which of the following factors?

    a. Stated goals

    b. Product potential

    c. Methodologies and measurements

    d. Optimal design

82. What is a numerical value that provides information about the entire system being studied?

    a. Parameter

    b. Statistic

    c. Population

    d. Sample

83. Those who have a vested interest in the process and/or its products and outputs are:

    a. subject matter experts.

    b. process owners.

    c. top management.

    d. stakeholders.

84. What is the main advantage of using focus groups to gather customer feedback?

    a. Rapid collection of large-scale quantitative data

    b. Anonymity of participant responses

    c. In-depth qualitative insights and group dynamics

    d. Automation of data analysis

85. Calculate the precision-to-tolerance (P/T) ratio using the following information. Specifications: 10 ± 0.5 cm and $\sigma_{total} = 0.045$ cm.

    a. 13.5%

    b. 27%

    c. 37%

    d. 75%

86. In a statistical test, power is defined as:

    a. the critical value of the test.

    b. Type II error.

    c. the probability of rejecting H_0 when it is false.

    d. the probability of accepting H_0 when it is true.

87. A factor that caused a nonconformance and should be permanently eliminated through process improvement is a(n):

    a. root cause.

    b. effect.

    c. dependent variable.

    d. critical value.

88. A *c*-chart tracks the:

    a. number of nonconforming units.

    b. rate of defects per unit.

    c. count of nonconformities.

    d. variation within subgroups.

89. Who can conduct an ISO 19011 third-party audit?

    a. The organization's director of quality

    b. A federal employee from a regulatory body

    c. An ASQ CQA from a neighboring plant

    d. A lead auditor working for a registrar

90. The requirements for a critical parameter being controlled would be located in which section of the control plan?

    a. Specification

    b. Characteristic

    c. Reaction plan

    d. Frequency

91. There are three positions at an IT help desk, each of which handles trouble tickets from a different department. How many ways can the 12 IT employees be assigned to the help desk?

    a. 4

    b. 12

    c. 220

    d. 1320

92. Using the interaction graph below, which factor settings will result in the largest mean response?

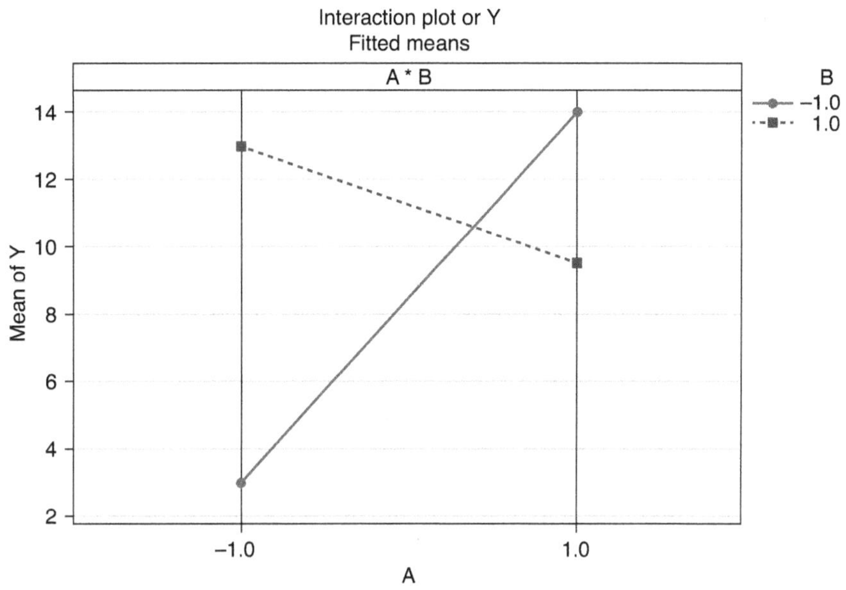

a. Factor A low, Factor B low

b. Factor A low, Factor B high

c. Factor A high, Factor B low

d. Factor A high, Factor B high

93. A team will perform a 5S in a supply room for the maintenance department. The first step is to:

a. install racks to hold equipment.

b. remove all obsolete equipment and materials.

c. designate a place for every piece of equipment.

d. clean the shelving and floor of the room.

94. Inventory, inspections, and rework are all forms of:

a. appraisal cost.

b. waste.

c. the value stream.

d. pull systems.

95. The measurement system analysis (MSA) term for variation in measurement when measured by two or more appraisers multiple times is:

    a. calibration.

    b. stability.

    c. repeatability.

    d. reproducibility.

96. Place the following flowcharting steps in order, from first to last:

    1. Build the sequence of the process, putting everything into the appropriate order.

    2. Determine the various steps in the process through team brainstorming or walking the process.

    3. Draw the flowchart, using the appropriate mapping symbols.

    4. Create the boundaries of the process of interest.

    5. Verify that the chart is complete.

    a. 2, 3, 4, 1, 5

    b. 4, 2, 1, 3, 5

    c. 2, 4, 1, 3, 5

    d. 4, 1, 3, 2, 5

97. In a multi-vari chart study, another term for *within-part* variation is:

    a. positional variation.

    b. cyclical variation.

    c. normal variation.

    d. temporal variation.

98. *np* is the mean of which distribution?

    a. Normal

    b. Chi-square

    c. Binomial

    d. Poisson

99. To determine the efficacy of a weight loss regimen, participants are weighed at the start and at the end of the 12-week program. Which pair of hypotheses would be most appropriate for this study?

a. H_0: $\mu_1 = \mu_2$ vs. H_1: $\mu_1 \neq \mu_2$

b. H_0: $\mu_1 - \mu_2 = 0$ vs. H_1: $\mu_1 - \mu_2 > 0$

c. H_0: $\mu_D = 0$ vs. H_1: $\mu_D \neq 0$

d. H_0: $\mu_D = 0$ vs. H_1: $\mu_D > 0$

100. A production team consists of seven workers. How many ways can two workers be chosen to serve on an improvement team?

a. 2

b. 7

c. 21

d. 42

SOLUTIONS

1. a; These are Deming's works. [I.A.1]
2. b; Brainstorming is a technique to elicit as many ideas as possible in a short period of time. [II.F.3]
3. b; The outcomes of jidoka, or autonomation, are automated stoppages and responses to quality issues and irregularities and detection of special cause variation. [VI.C.2]
4. c; The *p*-chart would be best for tracking the fraction of defective printed circuit boards produced each shift in which there is a varying sample size. [VI.A.3]
5. b; The RACI matrix is used to describe and define the roles and responsibilities of team members. [II.F.2]
6. c; Kaizen is a Japanese word meaning continuous improvement. [V.C.3]
7. a; A null hypothesis can only be rejected or fail to be rejected; it cannot be accepted because of a lack of evidence to reject it. [IV.B.2]
8. d; Check sheets use pre-categorized potential outcomes for data collection using sets of words, tally lists, or graphics. [III.D.2]
9. c; The Pareto chart is based on the 80/20 rule and helps teams narrow their focus down to the most important drivers of a problem. [II.C.3]
10. d; A scatter plot is the graphic tool best used for developing a complete picture of the relationship between two variables. [IV.A.2]
11. d; Defects per unit. Defects per million, rolled throughput yield, sigma levels, and capability are used to describe process performance. [II.E.1]
12. a; The likelihood that current controls will prevent a failure from reaching the customer is called detection. [I.C.3]
13. d; The symbol shown below is used to denote a supermarket in a value stream map. [I.B.2]

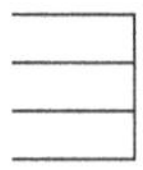

14. d; Also used as a risk tool, the C&E is not bound by the negative or positive and can be used to help understand the process. [II.D]
15. a; The 5S steps include sort, set in order, shine, standardize, and sustain. [I.B.2]
16. c; A half-fraction of a 2^6 experiment has $2^{6-1} = 2^5 = 32$ runs. [V.A.2]

17. a; The chi-square distribution is used to test a standard deviation or variance from a normal distribution. [IV.B.2]

18. b; Referring to an F table for $\alpha=0.01$, we find the value 4.51 corresponding to degrees of freedom $v_1=3$ and $v_2=30$. A useful reference is *The ASQ Certified Six Sigma Green Belt Handbook*, Appendix G. [III.C]

19. b; Mistake-proofing is achieved by limiting or restricting ways to complete a task to ensure accuracy and compliance. [V.C.1]

20. d; In the performing stage, team members realize that they are interdependent. The team focuses on the project goal. [II.F.1]

21. b; Setup time is defined as the time between the last piece of one product to the first good piece of another product, with the process running at full speed. [V.C.2]

22. d; The sources of variation included as part of an ANOVA table are between treatment, within treatment, and total. [IV.B.2]

23. a; A stem-and-leaf plot shows the shape, center, and spread of the data while still preserving the original data values. [III.D.4]

24. b; Using the table, we note that 140 defects were found last month. We then calculate the expected number of defects using the historical percentages as shown in the table below.

	Defect Type			
	Length	Weight	Strength	Total
Historical %	50%	25%	25%	100%
Observed inspection results	60	35	45	140
Expected	140(0.50) = 70	140(0.25) = 35	140(0.25) = 35	140

Noting that we have $k=3$ categories, we calculate the chi-square test statistic using

$$\begin{aligned}\chi_0^2 &= \frac{\sum_{i=1}^{k}(O_i - E_i)^2}{E_i} \\ &= \frac{(60-70)^2}{70} + \frac{(35-35)^2}{35} + \frac{(45-35)^2}{35} \\ &= 4.286.\end{aligned}$$

The chi-square critical value has $(k\text{-}1) = (3\text{–}1) = 2$ degrees of freedom. Using the chi-square table, we find $\chi^2(0.05, 2) = 5.991$. [IV.B.2]

Part B Solutions

25. b; If measurement were costly, it is likely that a sample size of 2 would be used for the *X*-bar and *R* chart. A sample size equal to 1 must be plotted using an *XmR* chart. [VI.A.3]

26. a; The voice of the customer uses warranty data, complaint logs, focus groups, and other customer-facing sources of data. [I.A.2]

27. b; Here, we will use the binomial distribution to solve since there is a constant probability of a defective, and the population size is not finite. Noting that $n = 5$ and $p = 0.15$, we solve $P(X = 1)$ using

$$P(X=1)=\binom{n}{x}p^x(1-p)^{n-x}$$

$$=\binom{5}{1}0.15^1(1-0.15)^{5-1}=5\times 0.15\times 0.85^4=0.3915.$$

[III.C]

28. d; Identifying and planning are usually connected with the start of a project; thus, the planning phase in a PDCA or PDSA. [II.C.6]

29. b; The RPNs are calculated using $S \times O \times D$ as shown in the table. RPNs, in order from highest to lowest, occur in rows 1, 3, 4, 2. [I.C.2]

Row	Severity	Occurrence	Detection	RPN
1	5	5	5	125
2	8	5	2	80
3	7	4	4	112
4	3	6	5	90

30. a; A common error made while performing a gage R&R study is altering the samples during the study. [III.E]

31. b; Response variables are the observed results of an experimental treatment. [V.A.1]

32. b; Training is not concluded until the imparted skills have been successfully adopted and demonstrated. This includes ongoing monitoring and verification of successful work completion, process adoption, and sustained improvements. [VI.B.3]

33. b; Cyclical variation is another term for part-to-part or lot-to-lot variation. [IV.A.1]

34. a; The weight variable follows a $N(16.0, 0.005^2)$ distribution. To find $P(X > 16.012)$, we will use the standard normal distribution.

$$P(X>16.012)=P\left(Z>\frac{X-\mu}{\sigma}\right)=P\left(Z>\frac{16.012-16.0}{0.005}\right)=P(Z>2.4)$$

Using a cumulative standard normal table, we find P(Z > 2.4) = 1 – P(Z < 2.4) = 1 – 0.9918 = 0.0082 = 0.82%. Note that the standard normal table in Appendix F of *The ASQ Certified Six Sigma Green Belt Handbook* gives values for P(Z > z), so the answer can be taken directly from the table. [III.C]

35. b; Visual devices do not require detailed text or content but rely on pictures, colors, and shapes. [VI.C.2]

36. d; To calculate the F statistic, we first calculate the mean square for the machine and error terms, as shown.

$$MS_{Machine} = \frac{SS_{Machine}}{df_{Machine}} = \frac{456.7}{4} = 114.2$$

$$MS_{Error} = \frac{SS_{Error}}{df_{Error}} = \frac{123.8}{24} = 5.158$$

$$F = \frac{MS_{Machine}}{MS_{Error}} = \frac{114.2}{5.158} = 22.13$$

Source	SS	DF	MSE	F
Machines	456.7	4	114.2	22.1341
Error	123.8	24	5.158	
Total	580.5	28		

[IV.B.2]

37. c; We compare the calculated F in the ANOVA table (see problem #36) with the critical value $F(4, 24, 0.05) = 2.78$. Since the calculated F is greater than the critical value, we reject H_0 and conclude that at least one machine mean is different from the rest. [IV.B.2]

38. b; To calculate the lower control limit for the *R* chart, we need to look up the table value D_3, which is based on subgroup size, *n*. Then, we use

$$LCL_R = D_3\bar{R}$$
$$D_3 \text{ for } n = 7 \text{ is } 0.076$$
$$LCL_R = 0.076 \times 16.25 = 1.235.$$

[VI.A.3]

39. d; The Poisson distribution is used to count defects per sampling unit. There is no natural upper limit to the number of defects found. [III.C]

40. d; Randomizing the order of the experimental runs will minimize the effect of uncontrolled variables on the response, thus increasing the statistical validity of the results. [V.A.1]

41. b; Fishbone, also known as cause-and-effect or Ishikawa, generally uses the 6Ms, which are manpower, materials, methods, Mother Nature, management, and measurement. Other input options include place, people, procedures, and policies. [IV.C.2]

42. c; The best choice here is multi-voting, the most common for decision-making. [II.F.3]

43. b; For all statistical tests, we reject the null hypothesis when the *p* value of the test is less than the alpha value. [IV.B1]

44. c; Upward flow is also known as bottom-up flow. [II.E.2]

45. a; In random sampling, each unit of the population has an equal chance of being chosen for the sample. [III.D.2]

46. a; Action steps to take when a critical parameter goes out of control would appear in the reaction plan column of a control plan. [VI.B.1]

47. c; Lean focuses on the value given to the customer versus the value given to your organization. [I.B.1]

48. c; The nominal group technique involves non-verbal interactions in which individual team members write down their brainstormed ideas first and then post them for the team to read. [II.F.3]

49. b; Defects per unit, or DPU, is calculated as the total number of defects/deficiencies divided by the total number of products produced. [II.E.1]

50. b; The result is statistically significant, but it is not practically significant based on the specifications. [IV.B.1]

51. b; In a full factorial experiment, all possible factor-level combinations are tested. [V.A.1]

52. b; A value stream map (VSM) is a detailed process map that shows all the steps in sequence in a value stream. Information gathered and displayed on a VSM can include cycle time, changeover time, number of people, WIP, inventory packaging sizes, scrap rates, value-added versus non-value-added activities, and number of changeovers. [I.B.2]

53. b; The median of a sample is the middle value of the sorted data list. By definition, the median is the second quartile, or the 50th percentile. [III.D.3]

54. c; The focus of quality function deployment is on customers and their satisfaction. [II.B.1]

55. b; Quick changeover, or single minute exchange of dies (SMED), is a technique that can significantly reduce the time taken to switch from producing one product to producing another. [V.C.2]

56. c; A negative correlation means that the two variables move in opposite directions. As x increases, y will tend to decrease. [IV.A.2]

57. d; The specification limits should reflect the voice of the customer. [VI.A.1]

58. c; For $x = \text{sales} = 45{,}678$ units, the expected number of returns is equal to:

$$\hat{y} = 230 + 0.0034x = 230 + 0.0034 \times 45{,}678 = 385.3. \text{ [IV.A.2]}$$

59. a; ISO 22301, security and resilience, business continuity management systems. [II.C.6]

60. a; A spaghetti diagram is a visual representation using a continuous flow line tracing the path of an item or activity through a process. [I.B.1]

61. c; To calculate the control limits for a *c*-chart, we first calculate the average number of defects found over the total time period, noting that data were collected over a total of $k = 25$ time periods:

$$\bar{c} = \frac{\sum_{i=1}^{k} c}{k} = \frac{38}{25} = 1.52.$$

Then,

$$LCL = \bar{c} - 3\sqrt{\bar{c}} = 1.52 - 3\sqrt{1.52} = -2.19 \uparrow 0$$
$$UCL = \bar{c} + 3\sqrt{\bar{c}} = 1.52 + 3\sqrt{1.52} = 5.22.$$

[VI.A.3]

62. a; Gage R&R refers to repeatability and reproducibility. [VI.B]

63. b; Most commonly, an A3 is used, but any one-to-two-page overview is what the champions and top leaders want to see. [II.C.6]

64. c; Using $s = 0.75$, $LSL = 44.0$, and $USL = 50.0$, we calculate

$$\hat{P}_p = \frac{(USL - LSL)}{6s} = \frac{(50.0 - 44.0)}{6 \times 0.75} = \frac{1}{0.75} = 1.33.$$

[III.F.3]

65. a; Based on the main effects plot, the mean response will decrease by 7.5 units (from 15 to 7.5) when Factor A is changed from the low to the high level. [V.A.2]

66. c; The roof of a house of quality displays the relationships between the technical requirements. [II.B.3]

67. d; According to the central limit theorem, the variance of the distribution of the sample averages is equal to the variance of the original distribution divided by n. [III.B.2]

68. c; The hypotheses for ANOVA are H_0: All means are equal vs. H_1: At least one mean is not equal to the rest. Since the p value of the F test is less than alpha, we reject H_0 and conclude at least one mean is not equal to the rest. [IV.B.2]

69. c; Restricting ways to complete a task to ensure accuracy and compliance is called poka-yoke, or mistake-proofing. [V.C.1]

70. b; Setting project boundaries can help the team distinguish between internal and external customers. [II.B.1]

71. b; Team formation discussion; the need to think about the project needs. [II.F.1]

72. a; A description of the direct relationship between process documents and the parent references for traceability. [VI.B.2]

73. d; Counts of events are classified as discrete or attribute data. [III.D.1]

74. b; Try-storming simulations can be used when information technology is applied to adjust or modify how a company receives, processes, and manages information. [V.B]

75. d; A blocking variable is not of experimental interest but is included in the analysis to reduce experimental error. [V.A.1]

76. d; SDCA is Standardize, Do, Check, Adjust. [VI.B]

77. a; By concentrating solely on speed, the process might be suboptimized if rework then increases. [II.C.4]

78. d; Pull systems are a lean tool used for waste elimination. [V.C.1]

79. b; Using the general multiplication rule, the probability of two events is calculated using $P(A \text{ and } B) = P(A) \times P(B|A)$. If Events A and B are independent, then $P(B|A) = P(B)$. Therefore, the multiplication formula reduces to $P(A \text{ and } B) = P(A) \times P(B)$. [III.B.1]

80. d; SMART goals can keep the team focused to avoid project scope creep. [I.A.2]

81. a; The project's stated goals are what the verification and validation (V&V) should focus on. Verification is assessment to specifications. Validation is demonstrated through actual usage in intended operating environments. [I.C.1]

82. a; The entire system being studied is the population. Populations are described using parameters. Samples are described using statistics. [III.B]

83. d; Interested parties is another name for stakeholders. [II.A.5]

84. c; Focus groups excel in providing in-depth qualitative insights by facilitating discussions among participants, allowing exploration of perceptions, opinions, and group dynamics. [II.D]

85. b; To calculate the precision-to-tolerance (P/T) ratio, we use

$$PTR = \frac{6\sigma_{total}}{(USL - LSL)} = \frac{6 \times 0.045}{(10.5 - 9.5)} = 0.27 = 27\%.$$

[III.E]

86. c; Power in a statistical test refers to the ability to detect a difference when it exists. Therefore, power is the probability of rejecting H_0 when it is false. [IV.B1]

87. a; This question asks how to prevent reoccurrences, thus requiring a root cause. [IV.C]

88. c; A *c*-chart tracks the count of nonconformities. Sample sizes are constant. [VI.A.3]

89. d; A lead auditor working for a registrar can conduct an ISO 19011 third-party audit. [VI.B.4]

90. a; The specification section of the control plan would provide the requirements for parameters being monitored. [VI.B.1]

91. d; Since the positions at the help desk are distinguishable, we must use a permutation to solve. We have $n = 12$ employees who can be assigned $r = 3$ distinct spots.

$${}_{12}P_3 = \frac{n!}{(n-r)!} = \frac{12!}{(12-3)!} = \frac{12!}{9!} = 12 \times 11 \times 10 = 1320$$

[III.B.1]

92. c; Setting Factor A to the high level and Factor B to the low level will result in a mean response equal to 14. [V.A.2]

93. b; The first step in 5S is to sort, which means to clear out any unnecessary equipment and materials. [V.C.1]

94. b; Inventory, inspections, and rework are all forms of waste. [V.C.1]

95. d; Reproducibility is the term for variation in measurement when readings are taken by two or more appraisers multiple times. [III.E]

96. b; The steps in creating a flowchart are to create the boundaries of the process of interest; determine the various steps in the process through team brainstorming or walking the process; build the sequence of the process, putting everything into the appropriate order; draw the flowchart, using the appropriate mapping symbols; and verify that the chart is complete. [III.A]

97. a; Positional variation is another term for within-part variation. [IV.A.1]

98. c; The mean of the binomial distribution is equal to the product of sample size and the probability of success, or *np*. [III.C]

99. d; Since each participant is weighed two times, this is an application of a paired sample test. We will conduct the test on the mean difference, μ_D. If the weight loss program is effective, the mean difference of (Before–After) weights will be positive. Therefore, H_0: $\mu_D = 0$ vs. H_1: $\mu_D > 0$ would be most appropriate for this study. [IV.B.2]

100. c; Since the workers are being chosen for a team, there is no inherent order. We can use the combination formula to solve:

$$nCr = 7C2 = \frac{n!}{r!(n-r)!} = \frac{7!}{2!(7-2)!} = \frac{7 \times 6}{2} = 21.$$

[III.B.1]

Appendix

The ASQ Certified Six Sigma Green Belt Body of Knowledge (2022)

Included in this Body of Knowledge (BoK) are explanations (subtext) and cognitive levels for each topic or subtopic in the test. These details will be used by the Examination Development Committee as guidelines for writing test questions and are designed to help candidates prepare for the exam by identifying specific content within each topic that can be tested. Except where specified, the subtext is not intended to limit the subject or be all-inclusive of what might be covered in an exam but is intended to clarify how topics are related to the role of the Certified Six Sigma Green Belt (SSGB). The descriptor in parentheses at the end of each subtext entry refers to the highest cognitive level at which the topic will be tested. A complete description of cognitive levels is provided at the end of this document.

I. Overview: Six Sigma and the Organization (11 Questions)

A. Six sigma and organizational goals

1. Value of six sigma

Recognize why organizations use six sigma, how they apply its philosophy and goals, and the evolution of six sigma from quality leaders such as Juran, Deming, Shewhart, Ishikawa, and others. (Understand)

2. Organizational goals and six sigma projects

Identify the linkages and supports that need to be established between a selected six sigma project and the organization's goals including SMART goals, and describe how process inputs, outputs, and feedback at all levels can influence the organization as a whole. (Understand)

3. Organizational drivers and metrics

Recognize key business drivers (profit, market share, customer satisfaction, efficiency, product differentiation, key performance indicators (KPIs)) for all types of organizations. Understand how key metrics and scorecards are developed and how they impact the entire organization. (Understand)

B. Lean principles in the organization

1. Lean concepts

Define and describe lean concepts such as theory of constraints, value chain, flow, takt time, just-in-time (JIT), Gemba, spaghetti diagrams, and perfection. (Apply)

2. Value-stream mapping

Use value-stream mapping to identify value-added processes and steps or processes that produce waste, including excess inventory, unused space, test inspection, rework, transportation, and storage. (Understand)

C. Design for six sigma (DFSS) methodologies

1. Road maps for DFSS

Distinguish between DMADV (define, measure, analyze, design, verify) and IDOV (identify, design, optimize, verify), and recognize how they align with DMAIC. Describe how these methodologies are used for improving the end product or process during the design (DFSS) phase. Understand how verification and validation are used to compare results against stated goals. (Understand)

2. Basic failure mode and effects analysis (FMEA)

Use FMEA to evaluate a process or product and determine what might cause it to fail and the effects that failure could have. Identify and use scale criteria, calculate the risk priority number (RPN), and analyze the results. (Analyze)

3. Design FMEA and process FMEA

Define and distinguish between these two uses of FMEA. (Apply)

II. Define Phase (20 Questions)

A. Project identification

1. Project selection

Describe the project selection process and what factors should be considered in deciding whether to use the six sigma DMAIC methodology or another problem-solving process. (Understand)

2. Process elements

Define and describe process components and boundaries. Recognize how processes cross various functional areas and the challenges that result for process improvement efforts. (Analyze)

3. Benchmarking

Understand various types of benchmarking, including competitive, collaborative, and best practices. (Understand)

4. Process inputs and outputs

Identify process input and output variables and evaluate their relationships using the supplier, inputs, process, output, customer (SIPOC) model. (Analyze)

5. Owners and stakeholders

Identify the process owners and other stakeholders in a project. (Apply)

B. Voice of the customer (VOC)

1. Customer identification

Identify the internal and external customers of a project, and what effect the project will have on them. (Apply)

2. Customer data

Collect feedback from customers using surveys, focus groups, interviews, and various forms of observation. Identify the key elements that make these tools effective. Review data collection questions to eliminate vagueness, ambiguity, and any unintended bias. (Apply)

3. Customer requirements

Use quality function deployment (QFD), Critical to X (CTX when 'X' can be quality, cost, safety, etc.), Critical to Quality tree (CTQ), and Kano model to translate customer requirements statements into product features, performance measures, or opportunities for improvement. Use weighting methods as needed to amplify the importance and urgency of different kinds of input; telephone call vs. survey response; product complaint vs. expedited service request. (Apply)

C. Project management basics

1. Project methodology

Define and apply agile and top-down project management methods. (Apply)

2. Project charter

Define and describe elements of a project charter and develop a problem statement that includes baseline data or current status to be improved and the project's goals. (Apply)

3. Project scope

Help define the scope of the project using process maps, Pareto charts, and other quality tools. (Apply)

4. Project metrics

Help develop primary metrics (reduce defect levels by x-amount) and consequential metrics (the negative effects that making the planned improvement might cause). (Apply)

5. Project planning tools

Use work breakdown structures (WBS), Gantt charts, critical path method (CPM), program evaluation and review technique (PERT) charts, and toll-gate reviews to plan projects and monitor their progress. (Apply)

6. Project documentation

Describe the types of data and input needed to document a project. Identify and help develop appropriate presentation tools (storyboards, spreadsheet summary of results) for phase reviews and management updates. (Apply)

7. Project risk analysis and management

Describe the elements of project risk analysis, including feasibility, potential impact, risk priority number (RPN), and risk management. Identify the potential effect risk can have on project goals and schedule, resources (materials and personnel), business continuity planning, costs and other financial measures, and stakeholders. (Understand)

8. Project closure

Review with team members and sponsors the project objectives achieved in relation to the charter and ensure that documentation is completed and stored appropriately. Identify lessons learned and inform other parts of the organization about opportunities for improvement. (Apply)

D. Management and planning tools

Define, select, and apply these tools: 1) affinity diagrams, 2) interrelationship digraphs, 3) tree diagrams, 4) prioritization matrices, 5) matrix diagrams, 6) process decision program charts (PDPC), 7) activity network diagrams, and 8) SWOT analysis. (Apply)

E. Business results for projects

1. Process performance

Calculate process performance metrics such as defects per unit (DPU), rolled throughput yield (RTY), cost of poor quality (COPQ), defects per million opportunities (DPMO), sigma levels, and process capability indices. Track process performance measures to drive project decisions. (Analyze)

2. Communication

Define and describe communication techniques used in organizations: top-down, bottom-up, and horizontal. (Apply)

F. Team dynamics and performance

1. Team stages and dynamics

Define and describe the stages of team evolution, including forming, storming, norming, performing, adjourning, and recognition. Identify and help resolve negative dynamics such as overbearing, dominant, or reluctant participants, the unquestioned acceptance of opinions as facts, groupthink, feuding, floundering, the rush to accomplishment, attribution, discounts, digressions, and tangents. (Understand)

2. Team roles and responsibilities

Use tools, such as RACI, to describe and define the roles and responsibilities of participants on Six Sigma and other teams, including black belt, master black belt, green belt, champion, executive, coach, facilitator, team member, sponsor, and process owner. (Apply)

3. Team tools and decision-making concepts

Define and apply team tools such as brainstorming, and decision-making concepts such as nominal group technique, and multi-voting. (Apply)

4. Team Communication

Identify and use appropriate communication methods (both within the team and from the team to various stakeholders) to report progress, conduct reviews, and support the overall success of the project. (Apply)

III. Measure Phase (20 Questions)

A. Process analysis and documentation

Develop process maps and review written procedures, work instructions, and flowcharts to identify any gaps or areas of the process that are misaligned. (Create)

B. Probability and statistics

1. Basic probability concepts

Identify and use basic probability concepts: independent events, mutually exclusive events, multiplication rules, permutations, and combinations. (Apply)

2. Central limit theorem

Define the central limit theorem and describe its significance in relation to confidence intervals, hypothesis testing, and control charts. (Understand)

C. Statistical distributions

Define and describe various distributions as they apply to statistical process control and probability: normal, binomial, Poisson, chi square, Student's t, and F. (Understand)

D. Collecting and summarizing data

1. Types of data and measurement scales

Identify and classify continuous (variables) and discrete (attributes) data. Describe and define nominal, ordinal, interval, and ratio measurement scales. (Analyze)

2. Sampling and data collection plans and methods

Define and apply various sampling methods (random and stratified) and data collection methods (check sheets and data coding). Prepare data collection plans that include gathering data and performing quality checks (e.g., minimum/maximum values, erroneous data, null values). (Apply)

3. Descriptive statistics

Define, calculate, and interpret measures of dispersion and central tendency. Develop and interpret frequency distributions and cumulative frequency distributions. (Evaluate)

4. Graphical methods

Construct and interpret diagrams and charts that are designed to communicate numerical analysis efficiently, including scatter diagrams, normal probability plots, histograms, stem-and-leaf plots, box-and-whisker plots. (Create)

E. Measurement system analysis (MSA)

Calculate, analyze, and interpret measurement system capability using gauge repeatability and reproducibility (GR&R) studies, measurement correlation, bias, linearity, percent agreement, and precision/tolerance (P/T). (Evaluate)

F. Process and performance capability

1. Process performance vs. process specifications

Define and distinguish between natural process limits and specification limits, and calculate process performance metrics. (Evaluate)

2. Process capability studies

Define, describe, and conduct process capability studies, including identifying characteristics, specifications, and tolerances, and verifying stability and normality. (Evaluate)

3. Process capability (C_p, C_{pk}) and process performance (P_p, P_{pk}) indices

Describe the relationship between these types of indices. Define, select, and calculate process capability and process performance. Describe when C_{pm} measures can be used. Calculate the sigma level of a process. (Evaluate)

4. Short-term vs. long-term capability and sigma shift

Describe the assumptions and conventions that are appropriate to use when only short-term data are used. Identify and calculate the sigma shift that occurs when long- and short-term data are compared. (Evaluate)

IV. Analyze Phase (18 Questions)

A. Exploratory data analysis

1. Multi-vari studies

Select appropriate sampling plans to create multi-vari study charts and interpret the results for positional, cyclical, and temporal variation. (Create)

2. Correlation and linear regression

Describe the difference between correlation and causation. Calculate the correlation coefficient and linear regression and interpret the results in terms of statistical significance (p-value). Use regression models for estimation and prediction. (Evaluate)

B. Hypothesis testing

1. Basics

Distinguish between statistical and practical significance. Determine appropriate sample sizes and develop tests for significance level, power, and type I and type II errors. (Apply)

2. Tests for means, variances, and proportions

Conduct hypothesis tests to compare means, variances, and proportions (paired-comparison t-test, F-test, analysis of variance (ANOVA), chi square) and interpret the results. (Analyze)

C. Additional analysis methods

1. Gap analysis

Analyze scenarios to identify performance gaps and compare current and future states using predefined metrics. (Analyze)

2. Root cause analysis

Use cause and effect diagrams, relational matrices, 5 Whys, fault tree analysis, and other problem-solving tools to identify the true cause of a problem. (Analyze)

V. Improve Phase (16 Questions)

A. Design of experiments (DOE)

1. Basic terms

Define and describe terms such as independent and dependent variables, factors and levels, responses, treatments, errors, repetition, blocks, randomization, effects, and replication. (Understand)

2. DOE graphs and plots

Interpret main effects analysis and interaction plots. (Apply)

B. Implementation Planning

Apply implementation planning by using proof of concepts, try-storming simulations, and pilot tests. (Apply)

C. Lean Tools

1. Waste elimination

Select and apply tools and techniques for eliminating or preventing waste, including pull systems, kanban, 5S, standard work, and poka-yoke. (Apply)

2. Cycle-time reduction

Use various techniques to reduce cycle time (continuous flow, setup reduction, single-minute exchange of dies (SMED)). (Analyze)

3. Kaizen and kaizen blitz

Define and distinguish between these two methods and apply them in various situations. (Apply)

VI. Control Phase (15 Questions)

A. Statistical process control (SPC)

1. SPC Basics

Describe the theory and objectives of SPC, including measuring and monitoring process performance for both continuous and discrete data. Define and distinguish between common and special cause variation and how these conditions can be deduced from control chart analysis. (Analyze)

2. Rational subgrouping

Define and describe how rational subgrouping is used. (Understand)

3. Control charts

Identify, select, construct, and use control charts: $\bar{X}$–R, $\bar{X}$–S, individual and moving range (ImR or XmR), median, p, np, c, and u. (Apply)

B. Sustain improvements

1. Control plan

Assist in developing and implementing a control plan to document and monitor the process. (Apply)

2. Document control

Understand document control and its role in controlling and sustaining improvements. (Understand)

3. Training plans

Develop training plans to implement and sustain improvements. (Apply)

4. Audits

Define first-, second-, and third-party audits. (Remember)

5. Plan-do-check-act (PDCA)

Apply and distinguish between the steps of plan-do-check-act (PDCA). (Apply)

C. Lean tools for process control

1. Total productive maintenance (TPM)

Define the elements of TPM, including use of predictive maintenance and describe how they can be used to control the improved process. (Understand)

2. Visual factory

Define the elements of a visual factory (Andon, Jidoka) and describe how they can be used to control the improved process. (Understand)

LEVELS OF COGNITION BASED ON BLOOM'S TAXONOMY—REVISED (2001)

In addition to **content** specifics, the subtext for each topic in this BOK also indicates the intended **complexity level** of the test questions for that topic. These levels are based on "Levels of Cognition" (from Bloom's Taxonomy—Revised, 2001) and are presented below in rank order, from least complex to most complex.

Remember

Recall or recognize terms, definitions, facts, ideas, materials, patterns, sequences, methods, principles, etc.

Understand

Read and understand descriptions, communications, reports, tables, diagrams, directions, regulations, etc.

Apply

Know when and how to use ideas, procedures, methods, formulas, principles, theories, etc.

Analyze

Break down information into its constituent parts and recognize their relationship to one another and how they are organized; identify sublevel factors or salient data from a complex scenario.

Evaluate

Make judgments about the value of proposed ideas, solutions, etc., by comparing the proposal to specific criteria or standards.

Create

Put parts or elements together in such a way as to reveal a pattern or structure not clearly there before; identify which data or information from a complex set is appropriate to examine further or from which supported conclusions can be drawn

www.ingramcontent.com/pod-product-compliance
Lightning Source LLC
LaVergne TN
LVHW061104130926
847902LV00030B/459